Li Shu-Chuan Cheng

MONOGRAPHS ON
STATISTICS AND APPLIED PROBABILITY

General Editors

**D.R. Cox, V. Isham, N. Keiding, N. Reid
and H. Tong**

(Full details concerning this series are available from the Publishers).

To Durgi, Ann and Pal

Variance Components Estimation

Mixed models, methodologies and applications

Poduri S.R.S. Rao

Professor of Statistics
University of Rochester, USA

CHAPMAN & HALL

London · Weinheim · New York · Tokyo · Melbourne · Madras

Published by Chapman & Hall, 2–6 Boundary Row, London SE1 8HN, UK

Chapman & Hall, 2–6 Boundary Row, London SE1 8HN, UK

Chapman & Hall GmbH, Pappelallee 3, 69469 Weinheim, Germany

Chapman & Hall USA., 115 Fifth Avenue, New York, NY 10003, USA

Chapman & Hall Japan, ITP-Japan, Kyowa Building, 3F, 2-2-1 Hirakawacho, Chiyoda-ku, Tokyo 102, Japan

Chapman & Hall Australia, 102 Dodds Street, South Melbourne, Victoria 3205, Australia

Chapman & Hall India, R. Seshadri, 32 Second Main Road, CIT East, Madras 600 035, India

First edition 1997

© 1997 Chapman & Hall

Printed in Great Britain by St Edmundsbury Press, Bury St Edmunds, Suffolk.

ISBN 0 412 728605

♾ Printed on permanent acid-free text paper, manufactured in accordance with ANSI/NISO Z39.48 - 1992 and ANSI/NISO Z39.48 - 1984 (Permanence of Paper).

Contents

Preface

From the very beginning of modern statistical methods more than 70 years ago, numerous articles appeared on several aspects of the variance components models. These models are widely used for statistical analysis and inference in agricultural, industrial and genetic research. They have also become essential in medical research and for ecological and environmental studies.

The Analysis of Variance (ANOVA) and Maximum Likelihood (ML) approaches were initially used for the estimation of the fixed effects and variance components of a mixed model and for the related statistical inference. The Minimum Norm Quadratic Unbiased Estimation (MINQUE) and the Minimum Variance Quadratic Unbiased Estimation (MIVQUE) with desirable properties offered fresh approaches to variance components estimation and established their relationship to the ML and the Restricted Maximum Likelihood Estimation (REML) procedures. The MINQUE and MIVQUE have also contributed to the emergence of new approaches to the nonnegative estimation of variance components, which remained elusive for a long time.

The purpose of this monograph is to comprehensively describe and illustrate the importance of the mixed models and present the different procedures available for estimating the variance and covariance components and the fixed effects of these models.

Starting with the historical origins of variance components models, the first chapter briefly describes the importance of these models for agricultural, genetic, medical and other types of research. Distributions of quadratic forms in normal variables, properties of the chi-square distribution and the approximations to the distributions of linear combinations of chi-square variables, which are essential for the study of variance components models, are presented in detail. This chapter ends with a list of references for related reading.

Estimation, hypothesis testing and confidence interval procedures for the one-way balanced and unbalanced models with fixed, random and mixed effects are presented in Chapter 2. These procedures for a ratio of variance components and the intraclass correlation coefficient are also presented. This chapter includes the Best Linear unbiased predictors (BLUP) for random effects and the Analysis of Covariance model with fixed and random effects.

Estimation of the variance components and fixed effects and the related statistical inference for the balanced and unbalanced two-way cross-classification are presented in Chapter 3, for the Randomized blocks, Balanced Incomplete Block Designs and Latin squares in Chapter 4, and the three-stage nested designs in Chapter 5.

In Chapters 2–5, the ANOVA method and its modifications are considered for the estimation of the variance components and fixed effects. The ML and REML approaches, including the iterative procedures, are presented in Chapter 6. The MINQUE and MIVQUE principles and their properties are studied in Chapter 7. Different procedures for nonnegative estimation of the variance components appear in Chapter 8.

The available approaches for the confidence intervals of the variance components and fixed effects are described in Chapter 9. Procedures for combining estimates from experiments or studies for meta-analysis and other purposes are presented in Chapter 10. The final chapter briefly describes the estimation of the variance and covariance components for the multivariate, nonlinear and generalized linear models and for repeated measures designs.

The different topics in all the chapters are illustrated through some of the studies which appeared in the literature and also with data from topics such as educational attainments, environmental pollution, ozone depletion, health care studies, hypertension and cholesterol levels. Exercises of both the theoretical and applied types are presented at the end of each chapter and their solutions appear at the end of the monograph. For the important procedures, alternative derivations are presented in the text or their verifications are suggested in the exercises.

This monograph can be recommended for a one semester course for graduate students in Statistics, Biostatistics and related areas. Graduate level courses in statistical estimation and inference, linear models and ANOVA procedures will provide the required preparation. Practitioners with familiarity in statistical methods will be able to use the different procedures presented in this monograph without any difficulty.

I am indebted to Dr. Ramana L. Rao for skillfully typesetting the entire monograph in LaTeX and for effortlessly implementing the frequent changes suggested by me. I would like to thank Mr. Mark Pollard for his editorial suggestions and Ms. Bea Shube for her initial interest in this project. Thanks to Bo Yang for finding the solutions to some of the examples and exercises through the SAS program and to Greg Elmer for proof reading some of the chapters. I would also like to thank my wife Durgi and my children Ann and Pal for their encouragement and patience during the preparation of this monograph.

Poduri S.R.S. Rao
Rochester, New York
March 1997

List of Tables

CHAPTER 1

Introduction

1.1 Historical origins and applications

Variance components are the different sources contributing to the variation in an observation. Fisher (1918, 1925) employed variance components analysis for his genetic research on Mendel's (1866) **laws of inheritance** and Darwin's (1824) hypothesis of **natural selection**. Subsequently, he laid the foundations for variance components models, designs of experiments, Analysis of Variance (ANOVA) techniques, statistical estimation and tests of hypotheses. Variance components models have been of importance in diverse fields of research and applications.

1.1.1 Genetic research and evolutionary studies

Various types of experiments are conducted in genetic laboratories and in green houses to study evolutionary changes and inheritance of different traits from one generation to the next. Variance components models containing both genetic and environmental effects are used for analyzing such experiments. Kempthorne (1963) and Falconer (1989) present genetic models and illustrations for studying the inheritance of quantitative traits.

1.1.2 Agricultural and industrial experiments

Agricultural experiments are conducted throughout the world for developing improved varieties of food grains, effective fertilizers and suitable cultivation methods. Horticultural experiments are conducted for increased productions of many flowers, plants, vegetables and seeds of desirable qualities. Similarly, experiments in animal husbandry, for instance, have increased milk yields. Variance components models are used for analyzing certain types of these experiments or for generalizing their results. These models are also extensively used for detecting the sources of variation in an industrial process and for improving the quality of production.

Some of the earliest contributions to the theory and applications of variance components models for industrial experiments were made by Tippett (1931) and Daniels (1939), and for agricultural experiments by Yates and Zacopanay (1935), Neyman *et al.* (1935), Cochran (1937, 1939), Yates and Cochran (1938), Yates (1940) and Winsor and Clarke (1940).

1.1.3 Medical research

Variance components analysis is used in medical research for more than one purpose. First, experimental designs are routinely employed in the research for producing medical compounds, and variance components models are used for analyzing some types of these designs. Secondly, a number of studies are conducted on patients and volunteers to ascertain the inheritance of hypertension, obesity, undesirable levels of cholesterol and similar factors affecting the health of individuals, and they are analyzed through suitable variance components models. Thirdly, cholesterol, glucose and other diagnostic measurements made by doctors have been found to vary with the measuring apparatus as well as the procedures employed for making the measurements. Standardizations of such diagnostic tests are attempted through appropriate experiments and suitable variance components analyses. Fourth, studies are conducted on the effects of smoking, alcohol consumptions, drug addiction, environmental pollution and similar factors on the health of the public. The patients, volunteers, laboratories and hospitals for these types of experiments and studies are usually considered to be random samples from the available sources and they are analyzed through suitable variance components models.

1.1.4 Further applications

Variance components models are also used for analyzing experimental and observational data in education, health, psychology and other social sciences. These models are widely employed in econometrics for analyzing data obtained from time series of cross-sectional units and for other types of research.

1.2 Classification of ANOVA models

A number of experimental designs have been developed to examine the **effects** of a set of **treatments** such as varieties of wheat, industrial processes and medical compounds. The ANOVA model describing the observations of an experimental design consists of the effects of the treatments, other classification factors of the design and the residual, the error. If all the effects of a model are nonrandom, it is known as the **fixed effects** model. On the other hand, if all the effects are considered to be random, it becomes the **random effects** model. Eisenhart (1947) classifies the fixed and random effects models as Models I and II respectively. The **mixed effects** model, or Model III, consists of both fixed and random effects. Models used for describing nonexperimental and observational data also follow this classification. Similarly, some of the coefficients of a linear or nonlinear model can also be fixed or random.

1.3 Quadratic forms in normal variables

1.3.1 Distribution of quadratic forms

The different sums of squares of a variance components model can be expressed as quadratic forms of the observations. Their distributions, expectations and variances, which are needed for the inferences on the variance components and fixed effects, are presented below.

1. Let Y denote the $n \times 1$ column vector of independent normal variables $y_i, i = 1, \ldots, n$, with zero means and unit standard deviations. A necessary and sufficient condition for the quadratic form $Q = Y'AY$ to have a χ^2-distribution is that A is idempo-

tent. The degrees of freedom (d.f.) of this distribution, denoted by f are equal to the rank of A. Since A is idempotent, f is equal to its trace (tr) , which is the sum of the diagonal elements.

2. The mean and variance of a χ^2-distribution with f d.f. are equal to f and $2f$ respectively.

3. If y_i have nonzero means μ_i and unit standard deviations, Q follows a noncentral χ^2-distribution with f d.f. and noncentrality parameter $\mu' A \mu$, where μ is the $n \times 1$ vector of the means.

4. Let Q_1 and Q_2 denote two independent χ^2 variables with f_1 and f_2 d.f. respectively. The ratio $(Q_1/f_1)/(Q_2/f_2)$ follows an F-distribution with f_1 and f_2 degrees of freedom.

5. If Y has a multivariate normal distribution with zero mean and covariance matrix Σ, the mean and variance of the quadratic form $Y'AY$ are equal to $tr A\Sigma$ and $2tr A\Sigma A\Sigma$ respectively.

1.3.2 Cochran's theorem (1934)

Consider, as above, n independent normal variables y_i with zero means and unit standard deviations. Let $Q_1 = Y'A_1Y$, $Q_2 = Y'A_2Y$, ..., $Q_k = Y'A_kY$ denote quadratic forms with ranks $n_1, n_2, \ldots, n_k$, and let

$$\sum_1^n y_i^2 = Q_1 + Q_2 + \ldots + Q_k . \tag{1.1}$$

(a) *Necessity.* If $Q_1, Q_2, \ldots, Q_k$ have independent χ^2-distributions with $n_1, n_2, \ldots, n_k$ d.f., $\sum_i y_i^2$ will have a χ^2-distribution with $n = \sum_i n_i$ d.f. For the proof, we note that the characteristic function of (1.1) is $E[e^{it(Q_1 + Q_2 + \ldots + Q_k)}] = (1 - 2it)^{n/2}$. Hence, $\sum_i y_i^2$ has a χ^2-distribution with n d.f.

(b) *Sufficiency.* If $\sum_i y_i^2$ has a χ^2-distribution, $Q_1, Q_2, \ldots, Q_k$ will have independent χ^2-distributions with $n_1, n_2, \ldots, n_k$ d.f. Cochran (1934) proves this result by considering an orthogonal transformation of Y.

This theorem also holds when y_i have independent normal distributions with means μ_i and unit standard deviations. In this case, $\sum_i y_i^2$ follows a noncentral χ^2-distribution with noncentrality parameter

$$\sum_{i=1}^{n} \mu_i^2 = \mu' A_1 \mu + \mu' A_2 \mu + \ldots + \mu' A_k \mu \ . \tag{1.2}$$

1.3.3 Standard errors of sample variances

Consider a random sample $y_i, i = 1, \ldots, n$, from a normal distribution with mean μ and variance σ^2. The sample mean $\bar{y} = \sum_i y_i / n$ and variance $s^2 = \sum_i (y_i - \bar{y})^2 / (n-1)$ are unbiased for μ and σ^2 respectively, and $(n-1)s^2/\sigma^2$ follows a χ^2-distribution with $(n-1)$ d.f. From this result or otherwise, the variance of s^2 is given by $2\sigma^4/(n-1)$

Let M denote a mean square. If $fM/E(M)$ is distributed as a χ^2-distribution with f d.f., the variance of M is $V(M) = 2E^2(M)/f$. Hence,

$$V(M) = E(M^2) - E^2(M) = E(M^2) - (f/2)V(M) \ . \tag{1.3}$$

Thus, $(f + 2)V(M)/2 = E(M^2)$ and an unbiased estimator of $V(M)$ is given by

$$v(M) = 2M^2/(f + 2) \ . \tag{1.4}$$

As an illustration, the estimator of the variance of s^2 is $2s^4/(n + 1)$.

1.3.4 Linear combinations of χ^2 variables

Consider the mean squares $M_i, i = 1, \ldots, k$, which have independent χ^2-distributions with f_i d.f, that is, $f_i M_i / E(M_i)$ have independent χ^2-distributions.

Estimators of variance components usually take the form of $M = \sum_i a_i M_i$, where a_i are constants. Following H.F.Smith (1938), Satterthwaite (1946) considers $fM/E(M)$ to be approximately distributed as a χ^2 with f d.f. As a consequence, $V(M) = 2E^2(M)/f$. However, $V(M) = \sum_i a_i^2 V(M_i), = 2 \sum_i [a_i^2 E^2(M_i) / f_i]$. Equating these two expressions for $V(M)$,

$$f = \frac{E^2(M)}{\sum_i [a_i^2 E^2(M_i)/f_i]} = \frac{[\sum_i a_i E(M_i)]^2}{\sum_i [a_i^2 E^2(M_i)/f_i]} \ . \tag{1.5}$$

In practice, f is obtained from $(\sum_i a_i M_i)^2 / \sum_i (a_i^2 M_i^2 / f_i)$.

1.4 Related reading

Cochran and Cox (1957), Cox (1958) and Kempthorne (1952) present the methodologies and applications related to the frequently used experimental designs. Anderson and Bancroft (1952), Brownlee (1965), Snedecor and Cochran (1967), Dunn and Clark (1974), John (1971), Mason, Gunst and Hess (1989), Milliken and Johnson (1992) and Montgomery (1990) among others contain some chapters on variance components analysis. C. R. Rao and Kleffé (1988) comprehensively present the estimation of variance and covariance components through quadratic forms of the observations. Sahai (1979), Sahai, Khuri and Kapadia (1985) and Khuri (1985) provide bibliographies on variance components models.

CHAPTER 2

One-way classification

2.1 Introduction

To examine the effect of a set of treatments, each of the treatments can be separately assigned to random samples of experimental units. This type of experiment is known as the completely randomized design. If the sample sizes for all the treatments are the same, the design is **balanced**; otherwise, it is **unbalanced**. The responses to the treatments from such a balanced or unbalanced design are analyzed through the **one-way fixed effects model** and the ANOVA procedure. This approach can also be used for examining the differences among the means of well-defined classes or groups.

In some situations, the treatments, classes or groups are considered to be selected randomly from large populations. In such cases, the above model becomes the **one-way random effects model**. The variances of the random effect and the residual, which are the variance components, can be estimated from the ANOVA procedure; alternative procedures for this purpose will be presented in Chapters 6–8.

The one-way random effects model is also used in some special types of applications. Cameron (1951) and Kusumal and Anderson (1967), for instance, describe the application of this model for analyzing **composite** samples, which are obtained from pooling observations of individual samples. Sukhatme and Narain (1984) consider this model with genetic effects to describe the daily energy balance of an individual. Kronberg, Hansen and Aaby (1992) use this model for analyzing data from measles epidemics. The applications of this model for combining information for **meta analysis** will be presented in Chapter 10.

Fisher (1918, 1925) introduced the **intraclass correlation** to measure the correlation between the members of the same family, group or class, and showed that it is given by the ratio of the variance of the random effect to the sum of the variances of the random effect and the residual of the one-way random effects model.

In some applications, **concomitant information** on the treatments becomes available, which is included in the one-way **analysis of covariance** model as one or more **covariates**. Through this model, the observations are adjusted for the covariates before testing the treatments of a fixed effects model or estimating the variance components of a random effects model.

2.2 Fixed effects and balanced designs

The observations for a one-way classification can be represented by the model

$$
\begin{aligned}
y_{ij} &= \mu_i + \varepsilon_{ij} \\
&= \mu + \alpha_i + \varepsilon_{ij} ,
\end{aligned}
\tag{2.1}
$$

where $\alpha_i = \mu_i - \mu$.

In this expression, $i = 1, 2, \ldots, k$, denotes the treatment, class or group and $j = 1, 2, \ldots, m$, denotes the sample observations. The samples are selected independently from each of the groups.

The mean of the ith group $E(y_{ij}|i)$ is denoted by μ_i and the overall mean of all the groups by μ. The deviation $\alpha_i = (\mu_i - \mu)$ is the **effect** of the ith treatment. The random error ε_{ij} is assumed to have mean zero and variance σ^2. The errors ε_{ij} and $\varepsilon_{ij'}(j \neq j')$ are assumed to be uncorrelated. Similarly, ε_{ij} and $\varepsilon_{i'j}$ of two groups are assumed to be uncorrelated.

2.2.1 Estimation of treatment effects

For the model in (2.1), the least squares estimators of μ and $\alpha_i, i = 1, 2, \ldots, k$ are obtained by minimizing

$$
\phi = \sum_{i=1}^{k} \sum_{j=1}^{m} (y_{ij} - \mu - \alpha_i)^2 .
\tag{2.2}
$$

From this approach, the **normal** equations are given by

$$\sum_{i=1}^{k}\sum_{j=1}^{m} y_{ij} = km\mu + m\sum_{i=1}^{k}\alpha_i \qquad (2.3)$$

and

$$\sum_{j=1}^{m} y_{ij} = m\mu + m\alpha_i. \qquad (2.4)$$

Let $\bar{y}_i = \sum_j y_{ij}/m$ and $\bar{y} = \sum_i \sum_j y_{ij}/n = \sum_i \bar{y}_i/k$ denote the mean of the ith group and the overall mean of the $n = km$ observations respectively. The $(k+1)$ equations in (2.3) and (2.4) are linearly dependent. For instance, the sum of the k equations in (2.4) is the same as (2.3). They can, however, be solved with a constraint such as one of the α_i or $\sum_i \alpha_i$ is zero. With the latter condition, which is frequently used in practice, the estimators of μ and α_i are given by $\hat{\mu} = \bar{y}$ and $\hat{\alpha}_i = (\bar{y}_i - \bar{y})$.

2.2.2 Inference regarding the means

The hypothesis that there is no difference among μ_i is the same as $\alpha_i = 0$ for $i = 1, 2, \ldots, k$. Through a general procedure employed for linear models, this hypothesis can be examined as follows. First, with the above estimators for μ and α_i, the minimum of (2.2) is given by

$$S_W = \sum_{i=1}^{k}\sum_{j=1}^{m}(y_{ij} - \hat{\mu} - \hat{\alpha}_i)^2 = \sum_{i=1}^{k}\sum_{j=1}^{m}(y_{ij} - \bar{y}_i)^2 . \qquad (2.5)$$

This is the **Within, Residual** or **Error** sum of squares (SS), which expresses the variation within the groups. Now, if the above hypothesis on α_i is valid, (2.2) is minimized when $\hat{\mu} = \bar{y}$. In this case, the minimum of (2.2) is given by $\sum_i \sum_j (y_{ij} - \bar{y})^2$, which is the **Total** SS. This sum of squares can be partitioned as

$$\sum_{i=1}^{k}\sum_{j=1}^{m}(y_{ij} - \bar{y})^2 = m\sum_{i=1}^{k}(\bar{y}_i - \bar{y})^2 + \sum_{i=1}^{k}\sum_{j=1}^{m}(y_{ij} - \bar{y}_i)^2 . \qquad (2.6)$$

The first term on the right hand side is the **Between** SS, S_B, which expresses the variation among the group means.

With the assumption of normality for y_{ij}, S_W/σ^2 follows a central χ^2-distribution with $k(m-1) = (n-k)$ d.f. Following the theorem in Section 1.3.2 or otherwise, S_B/σ^2 has a χ^2-distribution with $(k-1)$ d.f. independent of S_W/σ^2. This distribution is central if all the α_i are zero, and noncentral with noncentrality parameter $\sum_i \alpha_i^2/\sigma^2$ if at least one of the α_i is nonzero. From these results, a test for the above hypothesis on α_i is given by

$$F = M_B/M_W \; , \tag{2.7}$$

where $M_B = S_B/(k-1)$ and $M_W = S_W/(n-k)$ are the between and within mean squares (MSs) respectively. This ratio follows the F-distribution with $(k-1)$ and $(n-k)$ d.f.

Denoting the sample variance of the ith group by $s_i^2 = \sum_i (y_{ij} - \bar{y}_i)^2$, M_W can also be expressed as $s^2 = \sum_i s_i^2/k$, and it is unbiased for σ^2. The expected value of M_B is $\sigma^2 + m \sum_i \alpha_i^2/(k-1)$. Thus the expected value of the numerator of (2.7) becomes larger than that of the denominator if at least one of the means μ_i differs from the rest, that is, one of the α_i is nonzero.

If the variances of the treatments or groups are not the same, (2.7) does not follow the F-distribution, but an approximation to its distribution can be obtained from Section 1.3.4.

Example 2.1 Ozone depletion

To illustrate the model and the analysis in the above two sections, we examine the data on atmospheric ozone depletion. It is recognized that industrial pollutants such as chlorofluorocarbons have been rapidly depleting the protective ozone layer in the atmosphere. As a result, it is predicted that the Earth's environment and the ecosystem will be drastically disturbed in course of time. Specifically, there will be increased exposure to the Sun's ultraviolet rays, and temperatures will rise throughout the world. Consequently, plant and animal life will be damaged, and DNA structures of all forms of life may be permanently altered.

The *International Ozone Trends Panel Report (1990)* presents percentage changes in ozone abundance from 1965–1975 to 1976–1986 at 24 ground stations located in various places in the Northern Hemisphere. These ten year periods coincide with the solar cycles. Ozone depletion was observed at all the stations except one, more in winter from December through March than in summer from May to August.

Table 2.1. *Ozone depletion in winter: Percentage decreases from* 1965 − −1975 *to* 1976 − −1986

Latitude °N	19.5–39.3	40.0–47.8	50.2–64.1
	1.5	3.9	4.7
	0.3	2.9	3.2
	1.1	0.6	2.4
	1.7	1.3	4.7
	0.7	3.0	4.2
	1.8	2.8	3.8
	2.5	1.8	2.5
Sample size: m	7	7	7
Mean: $\bar{y}_i$	1.37	2.33	3.64
Variance: s_i^2	0.5424	1.2990	0.9362

Source: International Ozone Trends Panel Report (1990)

For 21 of the 24 stations, we have arranged in Table 2.1 the winter depletion rates into three groups of equal sizes. With some exceptions, the depletion rate is observed to increase with the latitude. The arrangement in the table corresponds to its increasing ranges.

For the station at 74.7 °N latitude which is above the Arctic Circle, "only less accurate moonlight measurements" were available during the winter. Only one station at 31.6 °N had a slight increase. We have excluded these two stations. Two stations at 46.8 °N have almost the same depletion rates and only one of them is included.

For the ozone data in Table 2.1, the Total, Between and Within sums of squares are 34.84, 18.18 and 16.66 respectively. From the last two figures, $M_B = 18.18/2 = 9.09$, $s^2 = M_W = 16.66/18 = 0.9256$, and $F = 9.09/.9256 = 9.82$ with 2 and 18 d.f. From the tables of the F-distribution, $F(2,18; .01) = 6.01$. Hence, the hypothesis of equality of the average depletion rates at the three latitude ranges can be rejected.

2.3 Random effects and balanced designs

In this case, the k treatments or groups are considered to be selected randomly from an infinite population. The observations from the groups can be represented by the model in (2.1) with the following description.

(a) The mean μ_i is a random variable with $E(\mu_i) = \mu$ and $V(\mu_i) = \sigma_\alpha^2$. (b) As a result, the treatment effect α_i is a random variable with mean zero and variance σ_α^2, and it is assumed to be uncorrelated with ε_{ij}. (c) The effects α_i and $\alpha_{i'}$, $i \neq i'$, of two treatments are assumed to be uncorrelated. (d) The assumptions for ε_{ij} are the same as those for the fixed effects case as described in Section (2.2). As a consequence of these assumptions,

1. $V(y_{ij}) = V(\alpha_i) + V(\varepsilon_{ij}) = \sigma_\alpha^2 + \sigma^2$,

2. $Cov(y_{ij}, y_{ij'}) = V(\alpha_i) = \sigma_\alpha^2$ for $j \neq j'$, and

3. $Cov(y_{ij}, y_{i'j}) = 0$ for $i \neq i'$.

The second result shows that the units in the ith group are correlated.

Estimation of the variance components $(\sigma_\alpha^2, \sigma^2)$ and the prediction of the random effect α_i is of chief interest in plant and animal improvement programs. Examining whether σ_α^2 is large or close to zero is usually of interest, especially in genetic applications. Estimation procedures for the above parameters and tests of hypotheses regarding them depend on whether the sample sizes for the groups are the same or unequal.

2.3.1 Estimation of the variance components

Expectations of M_B and M_W for both the fixed and random effects models are presented in Table 2.2. As noted in Section 2.2.2, the pooled variance $s^2 = \sum_i s_i^2 / k$ which is the same as M_W, is unbiased for σ^2. Since

$$E\left[\frac{\sum_i (\bar{y}_i - \bar{y})^2}{k - 1} \right] = \sigma_\alpha^2 + \sigma^2/m , \tag{2.8}$$

an unbiased estimator for σ_α^2 is

$$\hat{\sigma}_\alpha^2 = \frac{\sum_i (\bar{y}_i - \bar{y})^2}{k - 1} - \frac{s^2}{m} = \frac{M_B - M_W}{m} . \tag{2.9}$$

Table 2.2. *Expectation of mean squares (EMSs)*

Source	d.f.	MS	Fixed	Random
Between	$k-1$	M_B	$\sigma^2 + \frac{m}{k-1}\sum_i \alpha_i^2$	$\sigma^2 + m\sigma_\alpha^2$
Within	$n-k$	M_W	σ^2	σ^2

Note that an estimate obtained from this expression can be negative, in which case it is replaced by a small positive number.

If the three groups in Table 2.1 are considered to be a sample, the ozone depletion rates can be analyzed through the random effects model. For the data in this table, $s^2 = 0.9256$, $\sum_i(\bar{y}_i - \bar{y})^2/(k-1)$ $= M_B/m = 9.09/7 = 1.2986$, and hence $\hat{\sigma}_\alpha^2 = 1.2986 - 0.9256/7 = 1.17$.

2.3.2 Variances of the estimators

If y_{ij} has a normal distribution, $(n-k)s^2/\sigma^2$ follows a χ^2-distribution with $(n-k)$ d.f. and hence $V(s^2) = 2\sigma^4/(n-k)$. An unbiased estimator of this variance is given by $v(s^2) = 2s^4/(n-k+2)$.

Since $\bar{y}_i$ and s_i^2 are independent, the two terms on the right hand side of (2.6) are distributed independently. Furthermore, $\sum_i(\bar{y}_i - \bar{y})^2/(\sigma_\alpha^2 + \sigma^2/m)$ has a χ^2-distribution with $(k-1)$ d.f. From these results,

$$V(\hat{\sigma}_\alpha^2) = \frac{2}{k-1}\left(\sigma_\alpha^2 + \frac{\sigma^2}{m}\right)^2 + \frac{2}{(n-k)m^2}\sigma^4. \tag{2.10}$$

An unbiased estimator of this variance is given by

$$v(\hat{\sigma}_\alpha^2) = \frac{2}{k+1}\left[\frac{\sum_i(\bar{y}_i - \bar{y})^2}{k-1}\right]^2 + \frac{2\hat{\sigma}^4}{(n-k+2)m^2}$$

$$= \frac{2}{m}\left[\frac{M_B^2}{(n+m)} + \frac{M_W^2}{m(n-k+2)}\right]. \tag{2.11}$$

For the data in Table 2.1, $v(\hat{\sigma}_\alpha^2) = 0.85$ and hence the S.E. of $\hat{\sigma}_\alpha^2$ is 0.92.

2.3.3 Hypotheses regarding the variance components

The hypothesis that $\sigma^2 = \sigma_0^2$ against the alternative $\sigma^2 > \sigma_0^2$ can be tested with $(n-k)s^2/\sigma_0^2$, which follows the χ^2-distribution with $(n-k)$ d.f. Denoting the lower and upper $(1-\alpha)$ percentage points of the χ^2-distribution with $(n-k)$ d.f. by a and b, confidence limits for σ^2 are obtained from $(n-k)s^2/b$ and $(n-k)s^2/a$.

The hypothesis that $\sigma_\alpha^2 = 0$ against the alternative $\sigma_\alpha^2 > 0$ can be tested with $F = M_B/M_W$. This ratio follows the **central** F-distribution with $(k-1)$ and $(n-k)$ d.f. under both the null and alternative hypotheses.

For the ozone depletion example, as seen before, $F = 9.82$ is significant at the 0.01 level and the hypothesis that $\sigma_\alpha^2 = 0$ can be rejected.

2.3.4 Ratio of the variance components

As noted in Section 2.3.2,

$$U_1 = \frac{\sum_i (\bar{y}_i - \bar{y})^2}{\sigma_\alpha^2 + \sigma^2/m} = \frac{m \sum_i (\bar{y}_i - \bar{y})^2}{m\sigma_\alpha^2 + \sigma^2} = \frac{S_B}{m\sigma_\alpha^2 + \sigma^2} \tag{2.12}$$

and

$$U_2 = \frac{\sum_i \sum_j (y_{ij} - \bar{y}_i)^2}{\sigma^2} = \frac{S_W}{\sigma^2} \tag{2.13}$$

have independent χ^2-distributions with $(k-1)$ and $(n-k)$ d.f. respectively. The ratio

$$\mathcal{F} = \frac{U_1/(k-1)}{U_2/(n-k)} = \frac{1}{mR+1} \cdot \frac{M_B}{M_W} = \frac{F}{mR+1} , \tag{2.14}$$

where $R = \sigma_\alpha^2/\sigma^2$, has a **central** F-distribution with $(k-1)$ and $(n-k)$ d.f.

Let F_l and F_u denote the lower and upper $(1-\alpha)$ percentage points of the above distribution; that is,

$$P(F_l \leq \frac{F}{mR+1} \leq F_u) = (1-\alpha) . \tag{2.15}$$

From this result, $(1-\alpha)$ percent confidence interval for R is obtained from

$$\frac{1}{m}(\frac{F}{F_u} - 1) \leq R \leq \frac{1}{m}(\frac{F}{F_l} - 1) . \tag{2.16}$$

Approximate confidence limits for σ_α^2 may be obtained by multiplying the lower and upper limits in (2.16) by σ^2. Alternative procedures for finding these limits will be presented in Chapter 9.

The above ratio can be estimated from

$$\hat{R} = \hat{\sigma}_\alpha^2/\hat{\sigma}^2 = (M_B - M_W)/mM_W = (F - 1)/m \ . \tag{2.17}$$

The bias of this estimator becomes small as k or m becomes large and it can be completely eliminated; see Exercise 2.7.

2.3.5 Inference regarding the mean

The overall sample mean $\bar{y} = \sum_i \bar{y}_i/k$ is unbiased for μ and has variance

$$V(\bar{y}) = \frac{1}{k}(\sigma_\alpha^2 + \frac{\sigma^2}{m}) \ . \tag{2.18}$$

Since the expectation of $\sum_i(\bar{y}_i - \bar{y})^2/(k-1)$ is equal to $\sigma_\alpha^2 + \sigma^2/m$, an unbiased estimator of this variance is

$$v(\bar{y}) = \sum_i (\bar{y}_i - \bar{y})^2/k(k-1) \ . \tag{2.19}$$

The hypothesis that $\mu = \mu_0$ can be tested from

$$t_{k-1} = \frac{\bar{y} - \mu_0}{S.E.(\bar{y})} \ , \tag{2.20}$$

which follows the Student's t-distribution with $(k-1)$ d.f. The $(1-\alpha)$ confidence limits for μ are obtained from

$$\bar{y} - t.S.E.(\bar{y}), \qquad \bar{y} + t.S.E.(\bar{y}) \ . \tag{2.21}$$

The value of t is obtained from the t-distribution with $(k-1)$ d.f.

2.4 Intraclass correlation

In some studies, especially of the genetic type, it becomes important to study the relationship between the members of the same class, family, group or species. The intraclass correlation can be used for this purpose. With $i = 1, 2, \ldots, k$, denoting the class and $(j \neq j') = 1, 2, \ldots, m$, denoting the members of the same class, This correlation coefficient is defined as

$$\rho = \frac{Cov(y_{ij}, y_{ij'})}{\sqrt{V(y_{ij})V(y_{ij'})}} \ . \tag{2.22}$$

For the one-way random effects model, as noted in the beginning of Section 2.3, the covariance in the numerator is equal to σ_α^2 and $V(y_{ij}) = V(y_{ij'}) = \sigma_\alpha^2 + \sigma^2$. From these results,

$$\rho = \frac{\sigma_\alpha^2}{\sigma_\alpha^2 + \sigma^2} \ . \tag{2.23}$$

From the estimates for σ^2 and σ_α^2 in Section (2.3.1), an estimator for this correlation is

$$\hat{\rho} = \frac{M_B - M_W}{M_B + (m-1)M_W} = \frac{F-1}{F-1+m} \ , \tag{2.24}$$

where $F = M_B/M_W$ as before. Fisher (1925) defines the correlation in (2.23) and presents a large sample approximation to this estimator.

For a given Total SS, if the observations in the groups or the members in the classes are close to each other, the Within SS will be small relative to the Between SS. Consequently, a high value of $\hat{\rho}$ is associated with a large F. This result is also evident from (2.23). For the ozone data in Table 2.1, $\hat{\rho} = 0.56$.

2.5 Best linear unbiased predictor (BLUP)

To predict α_i, consider

$$\hat{\alpha}_i = \sum_{j=1}^{m} a_{ij} y_{ij} + b \ , \tag{2.25}$$

where a_{ij} and b are constants. For this predictor, $E(\hat{\alpha}_i - \alpha_i) = 0$ if $\mu \sum_j a_{ij} + b = 0$. Minimizing the MSE of α_i, $E(\hat{\alpha}_i - \alpha_i)^2$, with this condition,

$$\hat{\alpha}_i = \frac{\sigma_\alpha^2}{\sigma_\alpha^2 + \sigma^2/m}(\bar{y}_i - \mu) \ . \tag{2.26}$$

Now,

$$MSE(\hat{\alpha}_i) = \sigma_\alpha^2 \sigma^2 / (m\sigma_\alpha^2 + \sigma^2) \ . \tag{2.27}$$

Replacing μ in (2.26) by $\bar{y}$ and $(\sigma_\alpha^2, \sigma^2)$ by their estimators derived in Section 2.3.1,

$$\hat{\alpha}_i = (1 - \frac{M_W}{M_B})(\bar{y}_i - \bar{y}) \ . \tag{2.28}$$

Through the same procedure, the estimator for μ_i is given by

$$
\begin{aligned}
\hat{\mu}_i &= \frac{\sigma_\alpha^2}{\sigma_\alpha^2 + \sigma^2/m}\overline{y}_i + \frac{\sigma^2/m}{\sigma_\alpha^2 + \sigma^2/m}\mu \\
&= \overline{y}_i - \frac{\sigma^2/m}{\sigma_\alpha^2 + \sigma^2/m}(\overline{y}_i - \mu) .
\end{aligned}
\tag{2.29}
$$

The MSE of this estimator is the same as (2.27), which is smaller than σ^2/m. Note that if α_i are not assumed to be random, $V(\overline{y}_i) = \sigma^2/m$. Replacing $(\sigma_\alpha^2, \sigma^2; \mu)$ by their unbiased estimators,

$$
\hat{\mu}_i = \overline{y}_i - \frac{M_W}{M_B}(\overline{y}_i - \overline{y}) .
\tag{2.30}
$$

2.6 Fixed effects and unbalanced designs

For the one-way classification, the numbers of observations for the treatments or groups can be unequal in practical situations. Even if they are planned to be equal, missing observations and deletion of faulty measurements result in an unbalanced design. The model in (2.1) with $i = 1, 2, \ldots, k$, and $j = 1, 2, \ldots, n_i$, can represent the observations for this case. For the ANOVA method, μ is usually defined to be $\sum_i n_i\mu_i/n$. As a result, $\sum_i n_i\alpha_i = 0$.

The sample mean and variance for the ith group are $\overline{y}_i = \sum_i y_{ij}/n_i$ and $s_i^2 = \sum_i(y_{ij} - \overline{y}_i)^2/(n_i - 1)$. The overall mean of the $n = \sum_i n_i$ observations is $\overline{y} = \sum_i \sum_j y_{ij}/n = \sum_i n_i\overline{y}_i/n$. The Total SS can be expressed as

$$
\sum_{i=1}^{k}\sum_{j=1}^{n_i}(y_{ij} - \overline{y})^2 = \sum_{i=1}^{k} n_i(\overline{y}_i - \overline{y})^2 + \sum_{i=1}^{k}\sum_{j=1}^{n_i}(y_{ij} - \overline{y}_i)^2 .
\tag{2.31}
$$

The two terms on the right hand side are the Between SS (S_B) and Within SS (S_W) respectively.

With the assumption that the variance of ε_{ij} is the same for all the groups, an unbiased estimator of the common variance σ^2 is $\hat{\sigma}^2 = s^2 = \Sigma(n_i - 1)s_i^2/(n - k)$, which is the Within mean square M_W. The expectation of the Between mean square M_B is

$$
E\left[\frac{\sum_i n_i(\overline{y}_i - \overline{y})^2}{k - 1}\right] = \sigma^2 + \frac{\sum_i n_i\alpha_i^2}{k - 1} .
\tag{2.32}
$$

Table 2.3. *Ozone depletion rates in four regions*

Region	U.S.	Canada	Europe	Asia
	2.8	4.2	2.5	0.6
	3.0	4.7	3.8	0.7
	3.9	2.4	3.2	1.1
	1.8	1.3	4.7	0.3
	1.7		1.8	
	1.5		2.9	
			2.5	
Sample size: n_i	6	4	7	4
Mean: $\bar{y}_i$	2.45	3.15	3.06	0.675
Variance: s_i^2	0.883	2.50	0.9162	0.1092

The hypothesis of equality of the means can be tested from

$$F = \frac{\Sigma n_i (\bar{y}_i - \bar{y})^2 / (k-1)}{\Sigma\Sigma (y_{ij} - \bar{y}_i)^2 / (n-k)} = \frac{M_B}{M_W} \ . \tag{2.33}$$

As in the balanced case, this ratio follows the F-distribution with $(k-1)$ and $(n-k)$ d.f.

If the variances of ε_{ij} for the groups are not the same, (2.33) will not have the F-distribution but an approximation to its distribution can be found from Section 1.3.4.

Example 2.2 Ozone depletion in different regions

To illustrate the unbalanced design, in Table 2.3 we have arranged the ozone depletion rates into four regions: the U.S., Canada, Europe, and Asia.

The Total, Between and Within SS now are 34.87, 17.14, and 17.73 with 20, 3, and 17 d.f. respectively. From these figures, $M_B = 17.14/3 = 5.71$, $M_W = 17.73/17 = 1.04$, and F $= 5.71/1.04 = 5.49$. From tables of the F-distribution $F\,(3,17; 0.01) = 5.18$, which suggests that the average depletion rates are significantly different among the four geographical regions.

2.7 Random effects and unbalanced designs

Since ozone depletion rates can be measured at several regions around the world, the four regions in Table 2.3 can be considered to be a random sample from a large number of regions. In such a case, the data can be analyzed through the unbalanced model. The one-way unbalanced random effects model is widely used in a number of applications.

2.7.1 Estimation of the variance components

As seen in Section 2.3.1, $\hat{\sigma}^2 = \sum_i (n_i - 1)s_i^2/(n-k)$ is unbiased for σ^2. The expectation of S_B in (2.31) now is

$$E\left[\sum_i n_i(\bar{y}_i - \bar{y})^2\right] = \frac{\sum_i n_i(n - n_i)}{n}\sigma_\alpha^2 + (k-1)\sigma^2 \ . \qquad (2.34)$$

Substituting $\hat{\sigma}^2$ for σ^2, and removing the expectation sign on the left hand side, an unbiased estimator for σ_α^2 is

$$\begin{aligned}
\hat{\sigma}_\alpha^2 &= \frac{n}{\sum_i n_i(n-n_i)}\left[\sum_{i=1}^{k} n_i(\bar{y}_i - \bar{y})^2 - (k-1)\hat{\sigma}^2\right] \\
&= (M_B - M_W)/m_0 \ . \qquad\qquad (2.35)
\end{aligned}$$

where $m_0 = \sum_i n_i(n - n_i)/n(k-1)$. This estimator can take negative values and it becomes the same as (2.9) if $n_i \equiv m$.

For the data in Table 2.3,

$$\hat{\sigma}_\alpha^2 = \frac{21(3)}{324}(5.71 - 1.04) = 0.9081$$

2.7.2 Variances of the estimators

With the normality assumption for ε_{ij}, $(n-k)\hat{\sigma}^2/\sigma^2$ follows a χ^2-distribution with $(n-k)$ d.f., and hence $V(\hat{\sigma}^2) = 2\sigma^4/(n-k)$. Further, $\bar{y}_i$ and $\hat{\sigma}^2$ are independent, and from (2.35),

$$V(\hat{\sigma}_\alpha^2) = \frac{n^2\left[V(\sum_i n_i(\bar{y}_i - \bar{y})^2) + (k-1)^2 V(\hat{\sigma}^2)\right]}{(n^2 - \sum_i n_i^2)^2} \ . \qquad (2.36)$$

Let $\mathbf{Y}'$ denote the row vector with elements $(\bar{y}_i - \mu)$. The covariance matrix $\boldsymbol{\Sigma}$ of $\mathbf{Y}$ is diagonal with elements $V_i = V(\bar{y}_i) =$

$\sigma_\alpha^2 + \sigma^2/n_i$. Let $\mathbf{A}$ denote the $(k \times k)$ matrix with diagonal elements $a_{ii} = n_i(n - n_i)/n$ and off-diagonal elements $a_{ij} = -n_i n_j/n$. The Between SS can now be expressed as

$$\sum_i n_i(\bar{y}_i - \bar{y})^2 = \sum_i n_i[(\bar{y}_i - \mu) - (\bar{y} - \mu)]^2 = \mathbf{Y}'\mathbf{A}\mathbf{Y} . \quad (2.37)$$

From this result,

$$V\ \left[\sum_i n_i(\bar{y}_i - \bar{y})^2\right] = 2tr(\mathbf{A}\Sigma)^2$$

$$= \ 2\left[\sum_i n_i^2(1 - 2\frac{n_i}{n})V_i^2 + (\sum_i \frac{n_i^2}{n}V_i)^2\right] . \quad (2.38)$$

Now, from (2.36) and (2.38),

$$V\ (\hat{\sigma}_\alpha^2) = \frac{2n^2}{(n^2 - \sum_i n_i^2)^2}$$

$$\left[\sum_i n_i^2(1 - 2\frac{n_i}{n})V_i^2 + (\sum_i \frac{n_i^2}{n}V_i)^2 + \frac{(k-1)^2}{n-k}\sigma^4\right] . \ (2.39)$$

If n_i are all equal to m, this variance becomes the same as (2.10) for the balanced design. Crump, S.L. (1946, 1951) derived the estimator in (2.35) and its S.E. Anderson and Crump (1967) evaluated the performance of this estimator for different levels of unbalancedness.

2.7.3 Unweighted sums of squares (USS)

With the unweighted mean $\bar{y}_U = \sum_i \bar{y}_i/k$, an alternative unbiased estimator for σ_α^2 is

$$\hat{\sigma}_\alpha^2 = \frac{\sum_i(\bar{y}_i - \bar{y}_U)^2}{k - 1} - \frac{\hat{\sigma}^2}{m^*} , \quad (2.40)$$

where $m^* = k/\sum_i(1/n_i)$ is the harmonic mean of the n_i. The variance of this estimator is

$$V(\hat{\sigma}_\alpha^2) = 2\left[\frac{\sum_i V_i^2}{k(k-1)} + \frac{\sigma^4}{(n-k)m^{*2}}\right] . \quad (2.41)$$

For the data in Table 2.3, $\overline{y}_U = 2.33$ and $\sigma_\alpha^2 = 1.1084$ from (2.40). This estimate does not differ much from the ANOVA estimate of 0.9081 found earlier.

As can be expected, the variances in (2.39) and (2.41) for the ANOVA and USS estimators will be close to each other if n_i do not vary much. The relative magnitude of these variances in general depends on σ_α^2/σ^2. Further comparisons of these two estimators will be made in Chapter 7.

2.7.4 Ratio of the variance components and the intraclass correlation

For the unbalanced case, approximate confidence intervals for R can be obtained from (2.16) using the corresponding M_B and M_W and replacing m by m_0. For an alternative procedure, Thomas and Hultquist (1978) replace m by m^* and M_B by $m^* \sum_i (\overline{y}_i - \overline{y}_U)^2/(k-1)$.

An approximate estimator for ρ can be obtained as above from (2.24) by replacing m by m_0. C.A.B. Smith (1980) and others suggest alternative procedures of estimation. Harville and Fenech (1985) suggest a method for finding exact confidence intervals for R and hence for ρ. Burdick, Maqsood and Graybill (1986) present confidence intervals for ρ for the unbalanced one-way classification. Donner (1986) and Donner and Wells (1986) present a review of the procedures available for inferences regarding R and ρ.

2.7.5 Estimation of the mean

For μ, the weighted least squares estimator (WLS) $\overline{y}_W = \sum_i W_i \overline{y}_i / W$, where $W_i = 1/V_i$ and $W = \sum_i W_i$, is the minimum variance unbiased estimator with variance $V(\overline{y}_W) = 1/W$.

The sample mean $\overline{y}$ is unbiased for μ, and its variance $V(\overline{y}) = \sum_i (n_i/n)^2 V_i$ will be close to $V(\overline{y}_W)$ if σ_α^2 is close to zero.

The unweighted mean $\overline{y}_U$ is also unbiased for μ and its variance $V(\overline{y}_U) = \sum_i V_i/k^2$ approaches $V(\overline{y}_W)$ as σ_α^2/σ^2 becomes large.

The investigations of Cochran (1937, 1954) and Cochran and Carroll (1953) on the relative merits of $\overline{y}_U$ and $\overline{y}_W$ will be presented in Chapter 10.

2.8 Unequal residual variances and unbalanced designs

The sample variances for the data in Tables 2.1 and 2.3 indicate that the error variances σ_i^2 need not be equal. The model in (2.1) with $i = 1, 2, \ldots, k$, and $j = 1, 2, \ldots, n_i$, represents this general situation.

The expectation of S_B in (2.31) is

$$E\left[\sum_i n_i(\bar{y}_i - \bar{y})^2\right] = \frac{\sum_i n_i(n - n_i)}{n}\sigma_\alpha^2 + \frac{\sum_i (n - n_i)\sigma_i^2}{n} \tag{2.42}$$

Substituting s_i^2 for σ_i^2, an unbiased estimator for σ_α^2 is

$$\hat{\sigma}_\alpha^2 = \frac{n}{\sum_i n_i(n - n_i)}\left[\sum_{i=1}^{k} n_i(\bar{y}_i - \bar{y})^2 - \frac{\sum_i (n - n_i)s_i^2}{n}\right] . \tag{2.43}$$

For the data in Table 2.3,

$$\hat{\sigma}_\alpha^2 = \frac{21}{324}(17.14 - 3.35) = 0.8938 . \tag{2.44}$$

The variance of S_B is given by (2.37) with $V_i = \sigma_\alpha^2 + \sigma^2/n_i$, and $V(s_i^2) = 2\sigma_i^4/(n_i - 1)$. From these results,

$$\begin{aligned}
V(\hat{\sigma}_\alpha^2) &= \frac{2n^2}{[\sum_i n_i(n - n_i)]^2}\left[\sum_i n_i^2(1 - \frac{2n_i}{n})V_i^2\right. \\
&\quad + \left. (\sum_i \frac{n_i^2}{n}V_i)^2 + \sum_i (1 - \frac{n_i}{n})^2\frac{\sigma_i^4}{n_i - 1}\right] .
\end{aligned} \tag{2.45}$$

If σ_i^2 are not equal, the USS estimator for σ_α^2 is obtained by replacing the second term on the right hand side of (2.40) by $(1/k)\sum_i(s_i^2/n_i)$. The variance of $\hat{\sigma}_\alpha^2$ for this case is obtained from replacing the second term on the right hand side of (2.41) by $(1/k^2)\sum_i[\sigma_i^4/n_i^2(n_i^2 - 1)]$. Whether σ_i^2 are equal or unequal, the USS estimator for σ_α^2 can also take negative values. General procedures for nonnegative estimation of variance components will be examined in Chapter 8.

To estimate μ, $V_i = (\sigma_\alpha^2 + \sigma_i^2/n_i)$ for $\bar{y}_W$. As in the case of equal σ_i^2, $V(\bar{y})$ will be close to $V(\bar{y}_W)$ if σ_α^2 is close to zero. The variance of $\bar{y}_U$ will be close to $V(\bar{y}_W)$ if $\sigma_\alpha^2/\sigma_i^2$ are large.

2.9 Analysis of covariance

In a study on diets for weight reduction, the final weights of the experimental units can be expected to depend on their initial weights. Similarly, in a study on medical treatments, the responses of the experimental units are usually associated with their ages as well as some of the physiological characteristics. The initial weights, ages and similar factors are examples of **covariates** providing **concomitant** or supplementary information on the experimental units. The differences among the treatments can be examined after adjusting them with the observations on one or more covariates.

The fixed and mixed effects models with a single covariate are presented in the following sections, and they can be extended for the case of more than one covariate.

2.9.1 Fixed effects and balanced designs

The model in this case is

$$y_{ij} = \mu + \alpha_i + \beta(x_{ij} - \overline{x}) + \varepsilon_{ij} , \tag{2.46}$$

$i = 1, 2, \ldots, k$ and $j = 1, 2, \ldots, m$, where x_{ij} is the covariate and $\overline{x} = \sum_i \sum_j x_{ij}/n$ is its overall mean. As in Section 2.2, it is assumed that ε_{ij} is normally distributed with mean zero and variance σ^2.

The sums of squares in (2.6) can be expressed as $T_{yy} = B_{yy} + W_{yy}$. Similarly, the total sum of squares of x_{ij} can be expressed as $T_{xx} = B_{xx} + W_{xx}$. The sum of the cross-products (SP) is

$$\sum_{i=1}^{k}\sum_{j=1}^{m}(x_{ij} - \overline{x})(y_{ij} - \overline{y}) = m\sum_{i=1}^{k}(\overline{x}_i - \overline{x})(\overline{y}_i - \overline{y})$$

$$+ \sum_{i=1}^{k}\sum_{j=1}^{m}(x_{ij} - \overline{x}_i)(y_{ij} - \overline{y}_i) , \tag{2.47}$$

where $\overline{x}_i = \sum_j x_{ij}/m$ is the mean of the ith group. This equation can be expressed as $T_{xy} = B_{xy} + W_{xy}$.

For the model in (2.46), the least squares estimators with the condition $\sum_i \alpha_i = 0$ are given by $\hat{\mu} = \overline{y}$, $\hat{\beta} = W_{xy}/W_{xx}$ and $\hat{\alpha}_i = (\overline{y}_i - \overline{y}) - \hat{\beta}(\overline{x}_i - \overline{x})$. With these estimators, the Residual SS is

$$W_{y.x} = \sum_{i=1}^{k}\sum_{j=1}^{m}[(y_{ij} - \overline{y}_i) - \hat{\beta}(x_{ij} - \overline{x}_i)]^2$$

$$= W_{yy} - \hat{\beta}^2 W_{xx} = W_{yy} - W_{xy}^2/W_{xx} \ . \tag{2.48}$$

Note that W_{xy}^2/W_{xx} is the **Regression** SS. From this expression, $W_{y.x}/\sigma^2$ has a χ^2-distribution with $(n-k-1)$ d.f., and $W_{y.x}/(n-k-1)$ is an unbiased estimator for σ^2. Furthermore, $W_{xy}^2/W_{xx}\sigma^2$ has a χ^2-distribution with a single d.f. and it is independent of $W_{y.x}/\sigma^2$. From these results, the significance of β can be tested from

$$F = \frac{W_{xy}^2/W_{xx}}{W_{y.x}/(n-k-1)} \ , \tag{2.49}$$

which follows the F-distribution with 1 and $(n-k-1)$ d.f.

If the hypothesis that all the α_i are zero is valid, the least squares estimators of μ and β are $\overline{y}$ and T_{xy}/T_{xx} respectively. With these estimators, the residual SS is

$$T_{y.x} = T_{yy} - T_{xy}^2/T_{xx} \ . \tag{2.50}$$

In this case, T_{xy}^2/T_{xx} is the Regression SS, and $T_{y.x}/\sigma^2$ has a χ^2-distribution with $(n-2)$ d.f. From (2.48) and (2.50), $(T_{y.x} - W_{y.x})/\sigma^2$ has a χ^2-distribution with $(k-1)$ d.f. and it is independent of $W_{y.x}/\sigma^2$. Hence a test for $\alpha_i = 0$, $i = 1, 2, \ldots, k$, is given by

$$F = \frac{(T_{y.x} - W_{y.x})/(k-1)}{W_{y.x}/(n-k-1)} \ , \tag{2.51}$$

which follows the F-distribution with $(k-1)$ and $(n-k-1)$ d.f.

The adjusted estimator for $\mu_i - \mu_l = \alpha_i - \alpha_l, i \neq l$, is

$$(\hat{\mu}_i - \hat{\mu}_l) = (\overline{y}_i - \overline{y}_l) - \hat{\beta}(\overline{x}_i - \overline{x}_l) \ . \tag{2.52}$$

This estimator is unbiased and

$$V(\hat{\mu}_i - \hat{\mu}_l) = \left[\frac{2}{m} + \frac{(\overline{x}_i - \overline{x}_l)^2}{W_{xx}}\right]\sigma^2 \ . \tag{2.53}$$

To estimate this variance, σ^2 is replaced by its unbiased estimator $W_{y.x}/(n-k-1)$.

As can be seen from the tests in (2.49) and (2.51), and the adjustment in (2.46), it would be advantageous if the covariate is

Table 2.4. *Test score before (x) and after (y) enrolling in an educational program*

	Schools					
	1		2		3	
	x	y	x	y	x	y
	60	70	60	80	90	95
	65	70	80	90	85	90
	60	75	65	80	85	85
	70	85	85	90	80	90
	75	75	95	95	70	90
Means	66	75	77	87	82	90

highly related to the major variable of the study, and not affected by the treatments.

Example 2.3 Accelerated educational program

Test scores of samples of children from three schools before (x) and after (y) enrolling in an accelerated educational program are presented in Table 2.4. The sums of squares for x and y appear in columns three and four of Table 2.5, and the sums of products in column five.

Ignoring the covariate (x), for the y-scores, the between and within mean squares are $630/2{=}315$ and $\hat{\sigma}^2 = 380/12 = 31.67$, and hence $F = 315/31.67 = 9.95$ with 2 and 12 d.f. which is significant at the 0.01 level of significance. Estimates for the differences of the three means are $\bar{y}_2 - \bar{y}_1 = 12$, $\bar{y}_3 - \bar{y}_2 = 3$ and $\bar{y}_3 - \bar{y}_1 = 15$. The variance for each of these differences is $2\hat{\sigma}^2/m = 2(31.67)/5 = 12.67$, and hence the S.E. is 3.56.

With the covariate, the regression SS is $480^2/1230 = 187.32$ and the residual SS is $W_{y.x} = 380 - 187.32 = 192.68$. Hence, $\hat{\sigma}^2 = 192.68/11 = 17.52$. From (2.49), a test for $\beta = 0$ is $F = 187.32/17.52 = 10.69$ with 1 and 11 d.f., which is significant at the 0.01 level of significance.

Table 2.5. *Sums of squares and sums of products for test scores*

	d.f.	x	y	xy
Between	2	670	630	645
Within	12	1230	380	480
Total	14	1900	1010	1125

To test the means with the covariate, $T_{y.x} = 1010 - 1125^2/1900 = 343.88$ with $(n-2) = 13$ d.f., and $T_{y.x} - W_{y.x} = 151.20$ with $(k-1) = 2$ d.f. Hence, a test for the means is $F = 4.32$ with 2 and 11 d.f., which is significant at the 5 percent level of significance.

From (2.52), estimates of the differences of the means are $(\hat{\mu}_2 - \hat{\mu}_1) = 7.71$, $(\hat{\mu}_3 - \hat{\mu}_1) = 8.76$ and $(\hat{\mu}_3 - \hat{\mu}_2) = 1.05$. From (2.53), the S.E.s for these three differences are $2.95, 3.26$ and 2.71 respectively. Thus, in this illustration, the covariate adjustment has reduced the estimates for the differences of the means and also their S.E.s.

2.9.2 Fixed effects and unbalanced design

In this case, the model is given by (2.46) with $i = 1, 2, \ldots, k$ and $j = 1, 2, \ldots, n_i$. The group means are $\overline{x}_i = \sum_j x_{ij}/n_i$ and $\overline{y}_i = \sum_j y_{ij}/n_i$. The between SS and SP are given by

$$B_{xx} = \sum_{i=1}^{k} n_i(\overline{x}_i - \overline{x})^2 \, ,$$

$$B_{yy} = \sum_{i=1}^{k} n_i(\overline{y}_i - \overline{y})^2$$

and

$$B_{xy} = \sum_{i=1}^{k} n_i(\overline{x}_i - \overline{x})(\overline{y}_i - \overline{y}) \, . \tag{2.54}$$

The tests of hypotheses for β and μ_i take the same form as in Section 2.9.1. The variance in (2.53) becomes

$$V(\hat{\mu}_i - \hat{\mu}_l) = \left[\left(\frac{1}{n_i} + \frac{1}{n_l} \right) + \frac{(\overline{x}_i - \overline{x}_l)^2}{W_{xx}} \right] \sigma^2 \, . \tag{2.55}$$

Table 2.6. *Ozone depletion rates from 1965 to 1975 at three ground-based Antarctic stations. October means in Dobson Units. Estimated values are denoted by asterisks*

South Pole		Halley Bay		Syowa	
257.4	303.6	282.1	300.0	357.1	328.6
271.4	285.7	317.9*	301.1	378.6*	335.7
364.3	275.0	325.0	292.9	382.1	350.0
325.0	271.4	303.6	300.0*	335.7	321.4
260.7	253.6	282.1	310.7	292.9	285.7
300.0		282.1		332.1	

2.9.3 Random effects

Cochran (1946) considered the model in (2.46) with the assumption that α_i is normally distributed independent of ε_{ij} with mean zero and variance σ_α^2. In this case, the expected value of $W_{y.x}$ is $(n - k - 1)\sigma^2$ as before, and

$$E(T_{y.x}) = (n - 2)\sigma^2 + [(k - 2) + (W_{xx}/T_{xx})]m\sigma_\alpha^2 . \qquad (2.56)$$

From this result, an unbiased estimator for σ_α^2 is

$$\hat{\sigma}_\alpha^2 = \left[T_{y.x} - \frac{n - 2}{n - k - 1} W_{y.x} \right] \frac{(T_{xx}/m)}{(k - 2)T_{xx} + W_{xx}} . \qquad (2.57)$$

For the test scores in Example 2.3, we find the estimate for σ_α^2 from (2.57) to be 14.10. If the covariate is ignored, from the results in Section 2.3.1 and the fourth column of Table 2.5, we find that $\hat{\sigma}^2 = 31.67$ and $\hat{\sigma}_\alpha^2 = (315 - 31.67)/5 = 56.67$.

Exercises

2.1. Ozone depletion levels from 1965 to 1975 at three stations in the Antarctic are presented in Table 2.6. There is no apparent trend for this period. (a) From this data, estimate the variance components and the mean, and find the S.E.s of the estimates. (b) Find 95 percent confidence limits for σ^2, σ_α^2/σ^2 and μ.

2.2. The summer (May–August) percentage differences for 1976–

Table 2.7. *Ozone depletion in summer. Percentage decreases from 1965 −* *−1975 to 1976 − −1986*

	Latitude	
19.5 − 39.3	40.0 − 47.8	50.2 − 64.1
0.0	3.1	1.1
−0.4	−0.7	−1.2
−0.1	0.1	0.1
0.2	1.3	−0.8
0.5	1.4	1.4
3.3	0.6	0.9
0.7	−0.2	−1.7

1986 compared to 1965–1975 for the three latitude ranges of Table 2.1 are presented in Table 2.7. In Sections 2.2 and 2.3, we have used the model in (2.1) for the winter depletion rates. A similar model represents summer differences. The percentage changes, the random effects and the error terms for the winter and summer periods will be correlated. Let $d_{ij} = \mu_d + \alpha_{di} + \varepsilon_{di}$ represent the changes in the random percentage differences for these periods. (a) Estimate the variances $\sigma^2_{\alpha d}$ and σ^2_d of α_{di} and ε_{di}, and the mean μ_d. (b) Find the S.E.s of these three estimates. (c) Test the hypothesis that the rates of depletion are more in winter than in summer.

2.3. As can be seen from Table 2.3, ozone depletion rates for the first three regions are much higher than for the last region. These three regions can be considered to be a random sample from the industrialized nations. From the data of these three regions, estimate the variance components assuming that (a) σ^2_i are equal and (b) σ^2_i are unequal.

2.4. In an illustration presented by Cameron (1951), four samples of one-quarter pound of wool were tested from each of seven bales of wool. For the weights of the 'clean content', M_B and M_W were 10.9938 and 6.2606 respectively. Find the estimate of σ^2_α and its S.E.

Table 2.8. *Test scores for three schools*

	Schools					
	1		2		3	
	x	y	x	y	x	y
	50	65	70	85	80	85
	65	75	75	90	90	95
	55	70	70	90	85	95
	55	65	85	90	95	95
	50	70	80	90	90	90
Means	55	69	76	88	88	94

2.5. Test scores of children from three schools before (x) and after (y) completing an accelerated program of the type described in Example 2.3 are presented in Table 2.8. As in that example, test for the significance of the covariate, estimate the differences among the means with and without adjusting for the covariate and find their S.E.s.

2.6. For the data in Table 2.8, estimate σ_α^2 from (2.57).

2.7. (a) Noting that $E(1/M_W) = (n-k)/(n-k-2)\sigma^2$, show that $[(n-k-2)F/(n-k) - 1]/m$ is unbiased for R. (b) Show that the variance of this estimator is smaller than the MSE of $\hat{R}$.

2.8. Derive the predictor in (2.26) and the estimator in (2.29), and show that the minimum of the MSE is given by (2.27).

2.9. For an alternative method suggested by Snedecor and Cochran (1967, p.296) for the intraclass correlation, the numerator and denominator of (2.23) are estimated from the $km(m-1)$ pairs in the groups and the km units respectively. Express the estimator by this method in terms of M_B and M_W, and show that for large k, it becomes the same as (2.24).

2.10. For the unbalanced random effects model, S_B can be expressed as $\Sigma n_i(\overline{y}_i - \mu)^2 - n(\overline{y} - \mu)^2$. By finding the variances of these two terms and their covariance, show that the variance of S_B is the same as (2.38).

2.11. Find an approximation to the distribution of the estimator for the variance component in (2.9).

2.12 Following the approach of Section 2.3.4, find an exact confidence interval for the intraclass correlation coefficient.

2.13. In several industrial applications, the variance $T = \sigma_\alpha^2 + \sigma^2$ of y_{ij} is of importance. For the balanced design, find (a) an unbiased estimator $\hat{T}$ for T, (b) its variance $V(\hat{T})$, (c) an unbiased estimator for $V(\hat{T})$, and (d) an approximation to the distribution of $\hat{T}$.

2.14. Show that the expectation of $T_{y.x}$ is as presented in (2.56).

CHAPTER 3

Two-way cross-classification

3.1 Introduction

Some of the differences in the ozone depletions of the four regions in Table 2.1 can be attributed to the latitudes of the stations. Classifying the observations according to the regions and latitudes would enable the examination of the effects of both the factors on the depletion rates. Such an arrangement is an example of the two-way cross-classification.

For another illustration, consider an experiment to study the differences in the average gasoline mileages of three or four brands of automobiles. Samples of test runs can be used for examining these differences. However, factors such as speed, road and weather conditions may also effect the mileages. In a two-way classification scheme the mileages of different brands are examined at each of three or four levels of one of the other factors, for instance, speed. Higher order classifications can be considered to include the remaining factors.

Some of the observed differences among different types or brands of commercial products can be usually attributed to the differences among the skills and experiences of the persons actually manufacturing the products. Differences among the brands as well as the persons can be studied in a cross-classified design by assigning each person to each of the brands.

In these two-way designs, one or both of the classification factors can be **fixed** or **random**. There can also be an **interaction** between the factors. If the same number of observations are made for each of the different combinations of the two factors, the design or experiment is **balanced**; otherwise, it is **unbalanced**.

3.2 Fixed effects

The two-way cross-classification in general consists of rc cells formed by r rows and c columns. The observations can be represented by

$$y_{ijk} = \mu_{ij} + \varepsilon_{ijk}$$

$$= \mu + (\mu_{i.} - \mu) + (\mu_{.j} - \mu) + (\mu_{ij} - \mu_{i.} - \mu_{.j} + \mu) + \varepsilon_{ijk}$$

$$= \mu + \alpha_i + \beta_j + \gamma_{ij} + \varepsilon_{ijk} \tag{3.1}$$

for $i = 1, 2, \ldots, r$ rows, $j = 1, 2, \ldots, c$ columns, and $k = 1, 2, \ldots, m$ observations in the ijth cell. In this model, the overall mean is denoted by $\mu = \sum_i \sum_j \mu_{ij}/rc$. The row means and their effects are denoted by $\mu_{i.} = \sum_j \mu_{ij}/c$ and $\alpha_i = (\mu_{i.} - \mu)$. Similarly, $\mu_{.j} = \sum_i \mu_{ij}/r$ and $\beta_j = (\mu_{.j} - \mu)$ are the column means and their effects. With these definitions, $\sum_i \alpha_i = 0$ and $\sum_i \beta_j = 0$. The **interaction** between rows and columns is denoted by $\gamma_{ij} = (\mu_{ij} - \mu_{i.} - \mu_{.j} + \mu)$, and it will be explained further in Section 3.5. The random error ε_{ijk} is assumed to have mean zero and variance σ^2.

3.2.1 Additive model

Assuming that $\gamma_{ij} = 0$, (3.1) becomes the **additive** model

$$y_{ijk} = \mu + \alpha_i + \beta_j + \varepsilon_{ijk} \ . \tag{3.2}$$

In this case, the least squares estimators of μ, α_i and β_j are obtained by minimizing $\sum_{ijk}(y_{ijk} - \mu - \alpha_i - \beta_j)^2$.

The row and column means of the observations are $\bar{y}_{i..} = \sum_{jk} y_{ijk}/cm$ and $\bar{y}_{.j.} = \sum_{ik} y_{ijk}/rm$. The overall mean is $\bar{y} = \sum_{ijk} y_{ijk}/n = \sum_i \bar{y}_{i..}/r = \sum_j \bar{y}_{.j.}/c$, where $n = rcm$.

The Total SS can be expressed as

$$\sum_{ijk}(y_{ijk} - \bar{y})^2 = cm \sum_i (\bar{y}_{i..} - \bar{y})^2 + rm \sum_j (\bar{y}_{.j.} - \bar{y})^2$$

$$+ \sum_{ijk}(\bar{y}_{ijk} - \bar{y}_{i..} - \bar{y}_{.j.} + \bar{y})^2 \ . \tag{3.3}$$

The expressions on the right hand side are the sums of squares between rows and between columns, and the error or residual sum of squares. These sums of squares can be denoted by S_R, S_C and S_E respectively. With the assumption of normality for ε_{ijk}, S_R/σ^2,

Table 3.1. *Expected mean squares*

Source	MS	Fixed	Mixed	Random
Rows	M_R	$\sigma^2 + \frac{cm}{r-1}\sum_i \alpha_i^2$	$\sigma^2 + \frac{cm}{r-1}\sum_i \alpha_i^2$	$\sigma^2 + cm\sigma_\alpha^2$
Columns	M_C	$\sigma^2 + \frac{rm}{c-1}\sum_j \beta_j^2$	$\sigma^2 + rm\sigma_\beta^2$	$\sigma^2 + rm\sigma_\beta^2$
Residual	M_E	σ^2	σ^2	σ^2

S_C/σ^2, and S_E/σ^2 have independent χ^2-distributions with $(r-1)$, $(c-1)$ and $(n-r-c+1)$ d.f. respectively.

If the hypothesis of the equality of the row effects, that is, $\alpha_1 = \ldots = \alpha_r = 0$, is valid, S_R/σ^2 follows a central χ^2-distribution. Otherwise, it has a noncentral χ^2-distribution with noncentrality parameter $cm\sum_i \alpha_i^2/\sigma^2$.

Similarly, if the hypothesis of the equality of the column effects, that is, $\beta_1 = \ldots = \beta_c = 0$, is valid, S_C/σ^2 follows a central χ^2-distribution. Otherwise, it has a noncentral χ^2-distribution with noncentrality parameter $rm\sum_j \beta_j^2/\sigma^2$.

Let $M_R = S_R/(r-1)$, $M_C = S_C/(c-1)$, and $M_E = S_E/(n-r-c+1)$ denote the row, column and residual mean squares respectively. Expected values of these mean squares for the fixed effects case are presented in the third column of Table 3.1. The above hypotheses on the rows and columns can be examined from M_R/M_E and M_C/M_E. These ratios have F-distributions, with $(r-1)$ and $(c-1)$ d.f. respectively for the numerators. The denominator has $(n-r-c+1)$ d.f. in both cases.

Example 3.1 Love Canal samples

To illustrate the two-way cross-classification without interaction, we consider data from Love Canal which is situated in the town of Niagara Falls, and was used as a landfill for hazardous industrial waste. The U.S. Environmental Agency analyzed the migration of toxic material from this canal to the neighboring residential

Table 3.2. *Sample averages of three trace metals in five Love Canal site pairs. Amounts in milligrams per metric ton*

Site pairs	No. of samples	Chromium	Copper	Lead	Means
1	5	19.4	20.6	27.8	
11	13	20.2	18.9	20.6	21.3
2	15	17.0	17.6	18.5	
10	10	12.1	15.9	30.3	18.6
3	9	14.9	43.7	27.7	
8	9	34.6	17.7	35.8	29.1
4	15	25.6	19.8	25.9	
5	12	20.7	19.5	33.5	24.2
7	14	25.2	20.9	32.7	
9	10	25.2	21.7	40.8	27.8
Means		22.4	21.2	28.9	24.2

Source: Environmental Monitoring at Love Canal, Vol. 3, 1982.

area, which was divided into ten sites. In Table 3.2, we present the amounts of chromium, copper and lead found in samples selected from nine of the sites and the canal, which was designated as Site 11. These ten sites are grouped into five pairs based on their total amounts. The sixth site was not included since it had relatively larger amounts of copper and lead.

Ignoring the information in the second column on the numbers of samples, the ANOVA for the above data is presented in Table 3.3. From the F-ratios in this table, we find that the differences among the sites as well as the trace metals are significant; $F(4,23; 0.05) = 2.80$ and $F(2,23; 0.05) = 3.42$.

Table 3.3. *ANOVA of the Love Canal observations*

Source	d.f.	SS	MS	F
Between site pairs	4	460.3	115.08	2.95
Between trace metals	2	405.7	202.85	5.21
Residual	23	896.3	38.97	
Total	29	1762.3		

3.3 Mixed effects

The five site pairs of Table 3.2 can be considered to be a random sample from a large population of site pairs. To represent this situation through the model in (3.2) it is assumed that α_i is random with zero expectation and variance σ_α^2 and it is independent of ε_{ijk}. The column effects β_j for the three trace metals remain fixed with $\sum_j \beta_j = 0$. Expected values of the mean squares for this mixed model are presented in the fourth column of Table 3.1.

An unbiased estimator of σ_α^2 is

$$\hat{\sigma}_\alpha^2 = \frac{\sum_i (\bar{y}_{i..} - \bar{y})^2}{r-1} - \frac{\hat{\sigma}^2}{cm} = \frac{M_R - M_E}{cm} . \tag{3.4}$$

From this expression, the variance of $\hat{\sigma}_\alpha^2$ and its estimator are

$$V(\hat{\sigma}_\alpha^2) = \frac{2}{r-1}(\sigma_\alpha^2 + \frac{\sigma^2}{cm})^2 + \frac{2}{(cm)^2}\frac{\sigma^4}{n-r-c+1} \tag{3.5}$$

and

$$v(\hat{\sigma}_\alpha^2) = \frac{2}{(cm)^2}\left[\frac{M_R^2}{r+1} + \frac{M_E^2}{n-r-c+3}\right] . \tag{3.6}$$

From the mean squares for the Love Canal sites in Table 3.3,

$$\hat{\sigma}_\alpha^2 = (115.08 - 38.97)/6 = 12.7$$

and

$$v(\hat{\sigma}_\alpha^2) = \frac{2}{36}\left[\frac{115.08^2}{6} + \frac{38.97^2}{25}\right] = 126 \ .$$

Hence, S.E.$(\hat{\sigma}_\alpha^2) = 11.2$.

3.3.1 Inference for the mixed effects model

For the fixed effects, the test for $\beta_1 = \beta_2 = \ldots = \beta_c = 0$ was described in Section 3.2.1.

For testing the hypothesis that $\sigma_\alpha^2 = 0$, $\mathcal{F} = M_R/M_E$ follows the central F-distribution with $(r-1)$ and $(n-r-c+1)$ d.f. If this hypothesis is not valid, $[\sigma^2/(cm\sigma_\alpha^2 + \sigma^2)]\mathcal{F}$ has the central F-distribution with the same degrees of freedom.

The F-ratio of 2.95 in Table 3.3 for the sites indicates that the variation among the sites is significant; $F(4,23; 0.05) = 2.80$.

Denoting by F_l and F_u the $(1-\alpha)$ percentage points of the F-distribution with the above d.f., confidence limits for $R = \sigma_\alpha^2/\sigma^2$ are obtained from $(\mathcal{F} - F_u)/cmF_u$ and $(\mathcal{F} - F_l)/cmF_l$.

3.4 Both factors random

For the mixed model in the last section, the sites were considered to be a random sample. It may also be of interest to generalize the analysis of the data to trace metals of the type represented by the columns of Table 3.2. By the nature of the experiment or the data, in some cases both factors of a two-way classification may be considered to be random. These situations can be described by considering α_i and β_j of (3.2) to be random, with zero expectations and variances σ_α^2 and σ_β^2 respectively. Both these random effects and the errors ε_{ijk} are assumed to be uncorrelated. The expected mean squares for this case are presented in the last column of Table 3.1. Estimation of σ_α^2 and the inference regarding it remain the same as described in the last section.

An unbiased estimator for σ_β^2, its variance and unbiased estimator of variance are:

$$\hat{\sigma}_\beta^2 = \frac{\sum_j (\bar{y}_{.j.} - \bar{y})^2}{c-1} - \frac{\hat{\sigma}^2}{rm} = \frac{M_C - M_E}{rm}, \tag{3.7}$$

$$V(\hat{\sigma}_\beta^2) = \frac{2}{c-1}(\sigma_\beta^2 + \frac{\sigma^2}{rm}) + \frac{1}{(rm)^2}\frac{2\sigma^4}{n-r-c+1} \qquad (3.8)$$

and

$$v(\hat{\sigma}_\beta^2) = \frac{2}{(rm)^2}\left[\frac{M_C^2}{c+1} + \frac{M_E^2}{n-r-c+3}\right] . \qquad (3.9)$$

Tests for σ_α^2 and the confidence limits for σ_α^2/σ^2 were described in Section 3.3.1. Along the same lines, $\mathcal{F} = M_C/M_E$ provides a test for σ_β^2. This ratio follows a central F-distribution with $(c-1)$ and $(n-r-c+1)$ d.f. when $\sigma_\beta^2 = 0$. Otherwise, $[\sigma^2/(rm\sigma_\beta^2 + \sigma^2)]\mathcal{F}$ follows the same distribution. Confidence limits for σ_β^2/σ^2 are obtained from $(\mathcal{F} - F_u)/rmF_u$ and $(\mathcal{F} - F_l)/rmF_l$. The lower and upper percentage points F_l and F_u in this case are obtained from the F-distribution with the above degrees of freedom.

Example 3.2 Cholesterol measurements

It has been recognized that the measurements and analyses of blood samples of an individual often may vary with the measuring instruments, laboratories and analysts. Serious mistakes can occur in diagnosing a person's health from such results. Estimates of the variability arising from different sources can provide some guidelines in these situations.

As an illustration, we present the duplicate measurements of cholesterol levels made by a sample of three laboratories in Table 3.4 and their analysis in Table 3.5. Since the laboratories are considered to be a random sample, the column effects are assumed to be random. To generalize the results of this experiment to different cholesterol levels, the row effects are also assumed to be random.

The residual mean square 112.6 estimates σ^2. The sample variance of this estimator is $2(112.6)^2/20 = 1267.88$ and hence it has a S.E. of 35.6. From (3.4), $\hat{\sigma}_\alpha^2 = (5023.3 - 112.6)/6 = 818.4$ and from (3.6) it has a S.E. of 529.5. Similarly, $\hat{\sigma}_\beta^2 = (801.1 - 112.6)/8 = 86.1$ from (3.7), and from (3.9) it has a S.E. of 70.95.

The F-ratios of 44.6 and 7.1 are significant at a level of 0.001 or smaller, suggesting that σ_α^2 and σ_β^2 are large. Ninety percent confidence limits for σ_α^2/σ^2 and σ_β^2/σ^2 are (2.17, 64.47) and (0.12, 17.28) respectively.

Table 3.4. *Cholesterol levels (mg/100ml)*

		Laboratories			
		1	2	3	Means
		180	215	220	
	220	200	225	235	212.5
		220	230	235	
	240	240	235	255	235.8
Cholesterol levels		250	255	265	
	260	270	260	280	263.3
		265	270	280	
	280	275	285	290	277.5
Means		237.5	246.9	257.5	247.3

Table 3.5. *ANOVA for cholesterol measurements*

Source	d.f.	SS	MS	F
Between levels	3	15069.8	5023.3	44.6
Between laboratories	2	1602.1	801.1	7.1
Residual	18	2027.1	112.6	
Total	23	18699.0		

3.5 Fixed effects and interaction

A two way cross-classification can be used for studying, for instance, the differences among three or four methods of teaching in elementary schools. The teaching methods can be represented by the rows and the schools selected for the experiment by the columns. Performances of samples of students provide data for the experiment. If some of the schools had adopted some of the teaching methods before the experiment was conducted, it would have an effect on the performances of the students. In such a situation, there may be an **interaction** between the two classification factors.

A formal definition of the lack of interaction is that the difference between the means of any two rows is the same for every column, and the difference between the means of any two columns is the same for every row. From the first statement, $\mu_{ij}-\mu_{i'j} = \mu_{ij'}-\mu_{i'j'}$, for $i \neq i'$ and $j \neq j'$, that is, $\mu_{ij} - \mu_{i'j} - \mu_{ij'} + \mu_{i'j'}$ vanishes. Averaging this expression over (i',j'), $\gamma_{ij} = \mu_{ij} - \mu_{.j} - \mu_{i.} + \mu = 0$, where γ_{ij} is the interaction between rows and columns. The same result is obtained from the second statement. If this interaction exists, $\gamma_{ij} \neq 0$, and the model in (3.1) represents the observations. As before $\sum_i \alpha_i = 0$ and $\sum_j \beta_j = 0$. In addition, $\sum_i \gamma_{ij} = 0$ and $\sum_j \gamma_{ij} = 0$, and hence $\sum_{ij} \gamma_{ij} = 0$. For the least squares method, $\sum_i \sum_j (y_{ij} - \mu - \alpha_i - \beta_i - \gamma_{ij})^2$ is minimized with these constraints.

The row and column means, $\overline{y}_{i..}$ and $\overline{y}_{.j.}$, and the overall mean $\overline{y}$ are the same as defined in the previous sections. The mean of the ijth cell is $\overline{y}_{ij.} = \sum_k y_{ijk}/m$. The total sum of squares can be expressed as

$$\sum_{ijk}(y_{ijk} - \overline{y})^2 = \sum_{ijk}(y_{ijk} - \overline{y}_{ij.})^2 + m\sum_{ij}(\overline{y}_{ij.} - \overline{y})^2$$

$$= \sum_{ijk}(y_{ijk} - \overline{y}_{ij.})^2$$

$$+ m\sum_{ij}\left[(\overline{y}_{i..} - \overline{y}) + (\overline{y}_{.j.} - \overline{y}) + (\overline{y}_{ij.} - \overline{y}_{i..} - \overline{y}_{.j.} + \overline{y})\right]^2$$

$$= \sum_{ijk}(y_{ijk} - \overline{y}_{ij.})^2 + cm\sum_i(\overline{y}_{i..} - \overline{y})^2$$

$$+ \, rm \sum_j (\overline{y}_{.j.} - \overline{y})^2$$

$$+ \, m \sum_{ij} (\overline{y}_{ij.} - \overline{y}_{i..} - \overline{y}_{.j.} + \overline{y})^2. \qquad (3.10)$$

The first term on the right hand side is the sum of squares **within** the cells, or the error sum of squares S_E, with $rc(m-1) = (n-rc)$ d.f. The remaining terms are the sums of squares due to rows, columns, and interaction; S_R, S_C, and S_I. They have $(r-1)$, $(c-1)$, and $(r-1)(c-1)$ d.f. respectively.

The error mean square $M_E = S_E/(n-rc)$ is unbiased for σ^2. The row, column, interaction and error mean squares M_R, M_C, M_I, and M_E are obtained by dividing the respective sums of squares by the corresponding degrees of freedom. These mean squares and the tests for rows, columns, and interaction are presented in Table 3.6.

3.5.1 A test for additivity

If there is only one observation in each cell, the within sum of squares will be zero and the interaction cannot be tested as described above. Tukey (1949) considers the interaction to be of the multiplicative form $\lambda \alpha_i \beta_j$. An estimate for λ is obtained from the least squares method by substituting $(\overline{y}_{i.} - \overline{y})$ and $(\overline{y}_{.j} - \overline{y})$ for α_i and β_j respectively. If the hypothesis that $\lambda = 0$ is valid, the model for y_{ij} becomes **additive**, that is, $E(y_{ij}) = \mu + \alpha_i + \beta_j$.

The sum of squares for **nonadditivity** is obtained from

$$S_N = \frac{rc[\sum_i \sum_j (\overline{y}_{i.} - \overline{y})(\overline{y}_{.j} - \overline{y})y_{ij}]^2}{S_R S_C} \qquad (3.11)$$

which has 1 d.f. Hence the mean square for nonadditivity M_N is the same as S_N. Denoting the total SS by S_T, the residual sum of squares is given by

$$S_E = S_T - S_R - S_C - S_N , \qquad (3.12)$$

which has $(r-1)(c-1) - 1$ d.f. The residual mean square M_E is obtained by dividing (3.12) by these degrees of freedom. Tests for the rows, columns and nonadditivity are given by the F-ratios M_R/M_E, M_C/M_E and M_N/M_E respectively.

Table 3.6. *Expected mean squares (EMS) for the model with interaction; fixed effects*

Source	d.f.	MS	EMS	F
Rows	$r-1$	M_R	$\sigma^2 + cm\dfrac{\sum_i \alpha_i^2}{r-1}$	M_R/M_E
Columns	$c-1$	M_C	$\sigma^2 + rm\dfrac{\sum_j \beta_j^2}{c-1}$	M_C/M_E
Interaction	$(r-1)(c-1)$	M_I	$\sigma^2 + m\dfrac{\sum_i \sum_j \gamma_{ij}^2}{(r-1)(c-1)}$	M_I/M_E
Within	$rc(m-1)$	M_E	σ^2	
Total	$n-1$			

3.6 Random effects and interaction

If the row and column effects α_i and β_j in (3.1) are random, the interactions γ_{ij} will also be random. It is assumed that these three effects have zero expectations with variances σ_α^2, σ_β^2 and σ_γ^2 respectively. They are also assumed to be uncorrelated with each other and with the error term.

Expectations of the different mean squares obtained from (3.10) are presented in Table 3.7. From these expressions, an unbiased estimator for σ_α^2 and an estimator of its variance are

$$\hat{\sigma}_\alpha^2 = (M_R - M_I)/cm , \tag{3.13}$$

and

$$v(\hat{\sigma}_\alpha^2) = \frac{2}{(cm)^2}\left[\frac{M_R^2}{r+1} + \frac{M_I^2}{(r-1)(c-1)+2}\right] . \tag{3.14}$$

Similarly,

$$\hat{\sigma}_\beta^2 = (M_C - M_I)/rm \tag{3.15}$$

and

$$v(\hat{\sigma}_\beta^2) = \frac{2}{(rm)^2}\left[\frac{M_C^2}{c+1} + \frac{M_I^2}{(r-1)(c-1)+2}\right] . \tag{3.16}$$

Table 3.7. *Expected mean squares (EMS) for the model with interaction; random effects*

Source	d.f.	MS	EMS	F
Rows	$r - 1$	M_R	$\sigma^2 + cm\sigma_\alpha^2 + m\sigma_\gamma^2$	$\frac{M_R}{M_I}$
Columns	$c - 1$	M_C	$\sigma^2 + rm\sigma_\beta^2 + m\sigma_\gamma^2$	$\frac{M_C}{M_I}$
Interaction	$(r-1)(c-1)$	M_I	$\sigma^2 + m\sigma_\gamma^2$	$\frac{M_I}{M_E}$
Residual	$rc(m-1)$	M_E	σ^2	
Total	$n - 1$			

For the interaction,

$$\hat{\sigma}_\gamma^2 = (M_I - M_E)/m \tag{3.17}$$

and

$$v(\hat{\sigma}_\gamma^2) = \frac{2}{m^2}\left[\frac{M_I^2}{(r-1)(c-1)+2} + \frac{M_E^2}{n-rc+2}\right]. \tag{3.18}$$

An unbiased estimator for σ^2 is $\hat{\sigma}^2 = M_E$, and

$$v(\hat{\sigma}^2) = 2M_E^2/(n - rc + 2). \tag{3.19}$$

The hypothesis that $\sigma_\alpha^2 = 0$, $\sigma_\beta^2 = 0$ or $\sigma_\gamma^2 = 0$ can be tested from the F-ratios in Table 3.7. For the tests on σ_α^2 and σ_β^2, the interaction mean square M_I should be used in the denominator of the F-ratios. Note that for the fixed effects case, as shown in Table 3.6, the error mean square M_E appears in the denominator of the F-ratios. Johnson (1948) developed the tests of hypotheses for the balanced fixed and mixed effects models with interaction.

3.7 Mixed effects and interaction

For the illustration on the schools described in Section 3.5, the row effects α_i representing the teaching methods can be considered to be fixed with the assumption that $\sum_i \alpha_i = 0$. The columns should be considered to be random if the schools are selected randomly from a population of schools. In this case, β_j is assumed to have a normal distribution with mean zero and variance σ_β^2.

A number of practical situations can be represented by this **mixed effects** model. For instance, differences among a fixed number of specified apparatuses, machines, production processes and similar factors are frequently examined through industrial experimentation. The technicians assigned to these types of experiments are considered to be selected randomly.

Since the columns are random, the interaction effects γ_{ij} are also random. They are assumed to have normal distributions with zero means and variances $V(\gamma_{ij}|i)$. Since the rows are fixed, it is assumed that $\sum_i \gamma_{ij} = 0$, which implies that $\sum_i \sum_j \gamma_{ij} = 0$. This assumption also implies that (1) γ_{ij} and $\gamma_{i'j}$ for $i \neq i'$ are correlated, and (2) γ_{ij} and β_j are correlated; see Exercise 3.10.

Discussions on the above assumptions appear in Graybill (1961), Plackett (1960) and Scheffé (1956). Hocking (1973) points out the similarities between the different assumptions.

Since $\sum_i \alpha_i = 0$ and $\sum_i \gamma_{ij} = 0$,

$$\bar{y}_{i..} = \mu + \alpha_i + \beta + \gamma_{i.} + \bar{\varepsilon}_{i..},$$

$$\bar{y}_{.j.} = \mu + \beta_j + \bar{\varepsilon}_{.j.} \tag{3.20}$$

and

$$\bar{y} = \mu + \alpha_i + \beta + \bar{\varepsilon}, \tag{3.21}$$

where $\beta = \sum_j \beta_j / c$. Expected values of the different mean squares obtained from these expressions are presented in Table 3.8; $\sum_i V(\gamma_{ij}|i)/(r-1)$ is denoted by σ_γ^2.

The residual mean square M_E is unbiased for σ^2. Unbiased estimators for σ_β^2 and σ_γ^2 are

$$\hat{\sigma}_\beta^2 = (M_C - M_E)/rm \tag{3.22}$$

and

$$\hat{\sigma}_\gamma^2 = (M_I - M_E)/m . \tag{3.23}$$

A test for σ_β^2 is provided by the ratio M_C/M_E which has a central F-distribution with $(c-1)$ and $rc(m-1)$ d.f. if $\sigma_\beta^2 = 0$. If $\sigma_\beta^2 > 0$, $\sigma^2 M_C/(\sigma^2 + rm\sigma_\beta^2)M_E$ follows a central F-distribution with the same d.f.

For testing $\sigma_\gamma^2 = 0$, the ratio M_I/M_E follows the central F-distribution. If $V(\gamma_{ij}|i)$ is different for the rows, as pointed out by

Table 3.8. *Expected mean squares for the mixed model with interaction*

Source	d.f.	MS	EMS
Rows	$r-1$	M_R	$\sigma^2 + m\sigma_\gamma^2 + \frac{cm}{r-1}\sum_i \alpha_i^2$
Columns	$c-1$	M_C	$\sigma^2 + rm\sigma_\beta^2$
Interaction	$(r-1)(c-1)$	M_I	$\sigma^2 + m\sigma_\gamma^2$
Residual	$rc(m-1)$	M_E	σ^2
Total	$n-1$		

Scheffé (1959), $S_I/E(M_I)$ does not have a χ^2-distribution. In this case, it becomes complicated to find the power for $\sigma_\gamma^2 > 0$ from the above test. Approximations to the distribution of M_I/M_E are available in the literature.

If $V(\gamma_{ij}|i)$ is the same for all the rows, $\sigma_\gamma^2 = rV(\gamma_{ij}|i)/(r-1)$, and a test for $\alpha_1 = \alpha_2 = \ldots = \alpha_r = 0$ is given by M_R/M_I.

Wilk and Kempthorne (1955) consider the c levels of the second factor to be selected randomly from a population of C levels. As shown by them, for this case,

$$E(M_R) = \sigma^2 + cm\frac{\sum_i \alpha_i^2}{r-1} + m(1 - \frac{c}{C})\sigma_\gamma^2 \ . \tag{3.24}$$

If the population is large, this expression becomes the same as the expected mean square for rows presented in Table 3.8. If the column effects are fixed, $c = C$ and the last term vanishes.

3.8 Fixed effects and unbalanced designs

In some situations, the number of observations n_{ij} in the ijth cell can be unequal. Even when experiments are planned with the same n_{ij}, missing observations and deletion of outliers will make them unequal. When there is no interaction between rows and columns, such an unbalanced design can be represented by the model in

(3.2) with $i = 1, 2, \ldots, r$ rows, $j = 1, 2, \ldots, c$ columns and $k = 1, 2, \ldots, n_{ij}$ observations in the ijth cell.

Let $n_{i.} = \sum_j n_{ij}$, $n_{.j} = \sum_i n_{ij}$ and $n = \sum_i \sum_j n_{ij} = \sum_i n_{i.} = \sum_j n_{.j}$ denote respectively the total number of observations in the ith row, jth column and in the entire experiment. The totals of the ijth cell, ith row, and jth column respectively are $y_{ij.} = \sum_k y_{ijk}$, $y_{i..} = \sum_j \sum_k y_{ijk}$ and $y_{.j.} = \sum_i \sum_k y_{ijk}$. The corresponding means are $\overline{y}_{ij.} = y_{ij.}/n_{ij}$, $\overline{y}_{i..} = y_{i..}/n_{i.}$ and $\overline{y}_{.j.} = y_{.j.}/n_{.j.}$. The overall total and mean of all the n observations are $y_{...} = \sum_{ijk} y_{ijk}$ and $\overline{y} = y_{...}/n$.

Minimizing $\sum_{ijk}(y_{ijk} - \mu - \alpha_i - \beta_j)^2$, the estimators for $\hat{\mu}$, $\hat{\alpha}_i$, and $\hat{\beta}_j$ are obtained from

$$n\hat{\mu} + \sum_i n_{i.}\hat{\alpha}_i + \sum_j n_{.j}\hat{\beta}_j = y_{...} \; , \tag{3.25}$$

$$n_{i.}\hat{\mu} + n_{i.}\hat{\alpha}_i + \sum_j n_{ij}\hat{\beta}_j = y_{i..} \tag{3.26}$$

and

$$n_{.j}\hat{\mu} + \sum_i n_{ij}\hat{\alpha}_i + n_{.j}\hat{\beta}_j = y_{.j.} \; . \tag{3.27}$$

These $(r + c + 1)$ equations are linearly dependent. For instance, the sum of the r equations for the rows in (3.26) or the c equations for the columns in (3.27) is equal to (3.25). Thus, only $(r + c - 1)$ of these equations are linearly independent.

The estimators for α_i and β_j may be obtained with conditions such as $\sum_i n_{ij}\hat{\alpha}_i = 0$ and $\sum_j n_{ij}\hat{\beta}_j = 0$. These conditions imply that $\sum_i n_{i.}\hat{\alpha}_i = 0$ and $\sum_j n_{.j}\hat{\beta}_j = 0$, and the estimators are now given by $\hat{\alpha}_i = (\overline{y}_{i..} - \overline{y})$ and $\hat{\beta}_j = (\overline{y}_{.j.} - \overline{y})$. However,

$$Cov(\hat{\alpha}_i, \hat{\beta}_j) = -(\frac{n_{ij}}{n_{i.}n_{.j}} - \frac{1}{n})\sigma^2 \; , \tag{3.28}$$

which vanishes only if $n_{ij} = n_{i.}n_{.j}/n$, that is, n_{ij} should be proportional to the total sizes of the rows and columns. This condition, of course, is satisfied if n_{ij} are the same. Otherwise, inferences regarding α_i and β_j based on these correlated estimators can be misleading.

3.8.1 Estimation of the row and column effects

The row and column effects can be obtained from (3.25)–(3.27) without the assumptions in the last section.

To estimate α_i, from (3.27),

$$n_{i.}\hat{\mu} + \sum_j n_{ij}\hat{\beta}_j + \sum_j \left(\frac{n_{ij}}{n_{.j}} \sum_i n_{ij}\hat{\alpha}_i\right) = \sum_j n_{ij}\frac{y_{.j.}}{n_{.j}} . \tag{3.29}$$

Subtracting this equation from (3.26),

$$n_{i.}\hat{\alpha}_i - \sum_j \left(\frac{n_{ij}}{n_{.j}} \sum_i n_{ij}\hat{\alpha}_i\right) = y_{i..} - \sum_j n_{ij}\frac{y_{.j.}}{n_{.j}} = Q_i . \tag{3.30}$$

The right hand side expression is the **adjusted row total**. As can be seen, $\hat{\alpha}_i$ is obtained by first subtracting from $\overline{y}_{i..}$ the weighted average of the column means $\overline{y}_{.j.}$. The weights $n_{ij}/n_{i.}$ will all be equal to $1/c$ if n_{ij} for $j = 1, 2, \ldots, c$ are all the same.

The r equations in (3.30) are linearly dependent. For instance, the expressions on the left hand side add to zero. Estimators for α_i can be obtained with a constraint such as $\sum_i \alpha_i = 0$ or one of the α_is is zero.

To estimate β_j, from (3.26),

$$n_{.j}\hat{\mu} + \sum_i n_{ij}\hat{\alpha}_i + \sum_i \left(\frac{n_{ij}}{n_{i.}} \sum_j n_{ij}\hat{\beta}_j\right) = \sum_i n_{ij}\frac{y_{i..}}{n_{i.}} . \tag{3.31}$$

Subtracting this equation from (3.27),

$$n_{.j}\hat{\beta}_j - \sum_i \left(\frac{n_{ij}}{n_{i.}} \sum_j n_{ij}\hat{\beta}_j\right) = y_{.j.} - \sum_i n_{ij}\frac{y_{i..}}{n_{i.}} = Q'_j , \tag{3.32}$$

where the right hand side expression is the **adjusted column total**. Thus, $\hat{\beta}_j$ is obtained by subtracting from $\overline{y}_{.j.}$ the weighted average of the row means $\overline{y}_{i..}$. The weights $n_{ij}/n_{.j}$, will all be equal to $1/r$ if n_{ij} for $i = 1, 2, \ldots, r$ are all the same.

As in the case of the rows, the c equations in (3.32) are linearly dependent. Estimators for β_j can be obtained with a condition such as $\sum_j \beta_j = 0$ or one of the β_js is zero.

3.8.2 Inference regarding the row effects

To test the hypothesis that all the row effects are equal to zero, note that the residual sum of squares obtained from (3.25)–(3.27) is

$$
\begin{aligned}
S_E &= \sum_{ijk}(y_{ijk} - \hat{\mu} - \hat{\alpha}_i - \hat{\beta}_j)^2 \\
&= \sum_{ijk}(y_{ijk} - \overline{y})^2 - \Sigma\hat{\alpha}_i Q_i - \sum_j n_{.j}(\overline{y}_{.j.} - \overline{y})^2 .
\end{aligned}
\tag{3.33}
$$

The second expression is obtained by substituting $(\hat{\mu} + \hat{\beta}_j)$ from (3.27) in the first expression. The three terms in the second expression are the corrected total sum of squares, row sum of squares **adjusted** for columns $S_R(adj.)$, and column sum of squares **unadjusted** for rows, $S_C(unadj.)$. The distribution of S_E/σ^2 is χ^2 with $(n - r - c + 1)$ d.f. If the hypothesis that $\alpha_i = 0$ for $i = 1, 2, \ldots, r$ is valid, estimating μ and β_j from the least squares method, the residual sum of squares becomes

$$
S_H = \sum_{ijk}(y_{ijk} - \overline{y})^2 - \sum_j n_{.j}(\overline{y}_{.j.} - \overline{y})^2 .
\tag{3.34}
$$

From (3.33) and (3.34) the row SS **adjusted** for columns is $S_R(adj.)$ $= (S_H - S_E) = \sum_i \hat{\alpha}_i Q_i$.

Now, note that $S_R(adj.)/\sigma^2$ has a χ^2-distribution with $(r - 1)$ d.f. and it is independent of S_E/σ^2. The row and residual mean squares are $M_R = S_R(adj.)/(r - 1)$ and $M_E = S_E/(n - r - c + 1)$. A test for $\alpha_i = 0$ is provided by $F = M_R/M_E$ which follows the F-distribution with $(r - 1)$ and $(n - r - c + 1)$ d.f.

Example 3.3 Inter-laboratory assay of calcium

Brown, Healy and Kearns (1981) report on an assay of calcium in blood serum. The standard solutions with known and unknown concentrations were read on a spectrophotometer in arbitrary units. Four readings on the concentrations were obtained by each of twelve laboratories. For four of the solutions with *unknown* concentration levels U_1, U_2, Y_1 and Y_2 and three of the laboratories, we have obtained the unbalanced data in Table 3.9 by averaging

Table 3.9. *Calcium in blood serum*

Laboratories	U_1	U_2	Y_1	Y_2	$n_{i.}$	$\overline{y}_{i..}$
A	87	92	179	177	12	129.25
		84	83	173	182	
		80	76	166	172	
K	80	69	138	151	10	109.60
		70	46	138	138	
				132	134	
M	70	67	173	176	10	101.50
		60	63	166	148	
		44	48			
$n_{.j}$	8	8	8	8		
$\overline{y}_{.j.}$	71.875	68.0	158.125	159.75		

or deleting some of the four observations in each case; 1300 was subtracted from each of the observations. In this illustration, both the laboratories and the concentration levels can be considered to be random.

From the right hand side of (3.30), the adjusted row totals are $Q_1 = 1422/8$, $Q_2 = -1099/8$ and $Q_3 = -323/8$. From the left hand side of (3.30),

$$(60\hat{\alpha}_1 - 30\hat{\alpha}_2 - 30\hat{\alpha}_3)/8 = Q_1 ,$$

$$(-30\hat{\alpha}_1 + 54\hat{\alpha}_2 - 24\hat{\alpha}_3)/8 = Q_2$$

and

$$(-30\hat{\alpha}_1 - 24\hat{\alpha}_2 + 54\hat{\alpha}_3)/8 = Q_3 .$$

With $\hat{\alpha}_1 = 0$, we find that $\hat{\alpha}_2 = -28.6744$ and $\hat{\alpha}_3 = -18.7256$, and hence the adjusted row SS is $\hat{\alpha}_2 Q_2 + \hat{\alpha}_3 Q_3 = 4695.19$.

Similarly, with $\hat{\alpha}_2 = 0$, we find that $\hat{\alpha}_1 = 28.6744$ and $\hat{\alpha}_3 = 9.9487$, and hence the adjusted row SS, $\hat{\alpha}_1 Q_1 + \hat{\alpha}_3 Q_3$, again is 4695.19. The same row SS is obtained by solving the above equations with $\hat{\alpha}_3=0$.

Table 3.10. *ANOVA for laboratories with adjusted SS*

Source	d.f.	SS	MS	F
Laboratories adjusted	2	4695.19	2347.60	17.12
Concentrations unadjusted	3	63438.63		
Residual	26	3566.06	137.16	
Total	31	71699.88		

The remaining sums of squares and the test for the three laboratories are presented in Table 3.10. The F-ratio of 17.12 indicates that the differences among the laboratories is highly significant; $F(2,26; 0.001) = 9.12$.

3.8.3 Inference regarding the column effects

To test the hypothesis that all the β_j are equal to zero, consider the first expression of S_E in (3.33). Substituting for $(\hat{\mu} + \hat{\alpha}_i)$ from (3.26), this residual SS becomes

$$S_E = \sum_{ijk}(y_{ijk} - \overline{y})^2 - \sum_j \hat{\beta}_j Q'_j - \sum_i n_i.(\overline{y}_{i..} - \overline{y})^2 . \qquad (3.35)$$

The second and third terms on the right hand side are the **adjusted** column SS, $S_C(adj.)$ and unadjusted row SS, $S_R(unadj.)$ respectively. If the above hypothesis on the columns is valid, the residual sum of squares becomes

$$S'_H = \sum_{ijk}(y_{ijk} - \overline{y})^2 - \sum_i n_i.(\overline{y}_{i..} - \overline{y})^2 . \qquad (3.36)$$

Subtracting (3.35) from (3.36), $S_C(adj.) = (S'_H - S_E) = \sum_j \hat{\beta}_j Q'_j$.

Now, $S_C(adj.)/\sigma^2$ follows a χ^2-distribution with $(c - 1)$ d.f. and it is independent of S_E/σ^2. The hypothesis that $\beta_j = 0$ for

$j = 1, 2, \ldots, c$ can be tested from $F = M_C/M_E$, where $M_C = S_C(adj.)/(c-1)$. This ratio follows the F-distribution with $(c-1)$ and $(n - r - c + 1)$ d.f.

It can be seen from (3.33) and (3.35) that

$$\sum_i \hat{\alpha}_i Q_i + \sum_j n_{\cdot j}(\overline{y}_{\cdot j} - \overline{y})^2 = \sum_j \hat{\beta}_j Q_j + \sum_i n_{i\cdot}(\overline{y}_{i\cdot} - \overline{y})^2 . \tag{3.37}$$

Thus,

$$S_R(adj.) + S_C(unadj.) = S_C(adj.) + S_R(unadj.) . \tag{3.38}$$

Example 3.4 Interlaboratory assay of calcium

For the data in Table 3.9, with the overall mean of $\overline{y} = 114.44$, the unadjusted row SS is $\sum_i n_{i\cdot}(\overline{y}_{i\cdot} - \overline{y})^2 = 4540.725$. Now, from (3.37), the adjusted column SS is $S_C = 4695.19 + 63438.625 - 4540.725 = 63593.09$, and hence $M_C = 63593.09/3 = 21197.70$. The F-ratio, $21197.70/137.16 = 154.6$, indicates that the differences among the concentrations are significant; $F(3, 26; 0.001) = 7.36$.

3.8.4 Unbalanced designs with interaction

The model in (3.1) with $k = 1, 2, \ldots, n_{ij}$ observations in the ijth cell represents this design. The total sum of squares can be expressed as

$$\sum_{ijk}(y_{ijk} - \overline{y})^2 = \sum_{ijk}(y_{ijk} - \overline{y}_{ij\cdot})^2 + \sum_{ij} n_{ij}(\overline{y}_{ij\cdot} - \overline{y})^2 . \tag{3.39}$$

The right hand side expressions are the within cell and between cell sums of squares, S_W and S_B, respectively. We note that S_W/σ^2 and S_B/σ^2 have χ^2-distributions with $(n-rc)$ and $(rc-1)$ d.f. The adjusted and unadjusted row and column sums of squares remain the same as in the previous two sections. The interaction sum of squares is

$$\begin{aligned}
S_I &= S_B - S_R(adj.) - S_C(unadj.) \\
&= S_B - S_R(unadj.) - S_C(adj.) .
\end{aligned} \tag{3.40}$$

Let $M_I = S_I/(r-1)(c-1)$ and $M_W = S_W/(n-rc)$ denote the interaction and within cell mean squares. The F-ratios M_R/M_W,

M_C/M_W and M_I/M_W provide the tests for rows, columns and interaction.

3.9 Fixed effects and proportional frequencies

If $n_{ij} = n_{i.}n_{.j}/n$, the least squares equations (3.26) and (3.27) for the model in (3.2) become

$$\hat{\mu} + \hat{\alpha}_i + \frac{1}{n}\sum_j n_{.j}\hat{\beta}_j = \overline{y}_{i..} \tag{3.41}$$

and

$$\hat{\mu} + \frac{1}{n}\sum_i n_{i.}\hat{\alpha}_i + \hat{\beta}_j = \overline{y}_{.j.} \ . \tag{3.42}$$

As noted in Section 3.8, the conditions $\sum_i n_{ij}\hat{\alpha}_i = 0$ and $\sum_j n_{.j}\hat{\beta}_j = 0$ imply that $\sum_i n_{i.}\hat{\alpha}_i = 0$ and $\sum_j n_{.j}\hat{\beta}_j = 0$. From (3.25) and the above two equations, $\hat{\alpha}_i = (\overline{y}_{i..} - \overline{y})$ and $\hat{\beta}_j = (\overline{y}_{.j.} - \overline{y})$

The total SS in this case can be expressed as

$$\sum_{ijk}(y_{ijk} - \overline{y})^2 = \sum_i n_{i.}(\overline{y}_{i..} - \overline{y})^2 + \sum_j n_{.j}(\overline{y}_{.j.} - \overline{y})^2$$

$$+ \sum_{ijk}(y_{ijk} - \overline{y}_{i..} - \overline{y}_{.j.} + \overline{y})^2 \ . \tag{3.43}$$

The expressions on the right hand side are the row, column and error sums of squares, S_R, S_C and S_E respectively. The expected values of the corresponding mean squares M_R, M_C and M_E are presented in Table 3.11. The F-ratios M_R/M_E and M_C/M_E provide the tests for $\alpha_i = 0$ for $i = 1, 2, \ldots, r$ and $\beta_j = 0$ for $j = 1, 2, \ldots, c$ respectively.

Similarly, for the model with interaction in (3.1) ,

$$\sum_{ijk}(y_{ijk} - \overline{y})^2 = \sum_i n_{i.}(\overline{y}_{i..} - \overline{y})^2 + \sum_j n_{.j}(\overline{y}_{.j.} - \overline{y})^2$$

$$+ \sum_{ij} n_{ij}(\overline{y}_{ij.} - \overline{y}_{i..} - \overline{y}_{.j.} - \overline{y})^2$$

$$+ \sum_{ijk}(y_{ijk} - \overline{y}_{ij.})^2 \ . \tag{3.44}$$

Table 3.11. *Expected mean squares for proportional frequencies*

Source	d.f.	MS	EMS
Rows	$r-1$	M_R	$\sigma^2 + \dfrac{\sum_i n_{i.}\alpha_i^2}{r-1}$
Columns	$c-1$	M_C	$\sigma^2 + \dfrac{\sum_j n_{.j}\beta_j^2}{c-1}$
Error	$n-r-c+1$	M_E	σ^2
Total	$n-1$		

Table 3.12. *Expected mean squares for proportional frequencies and interaction; fixed effects*

Source	d.f.	MS	EMS
Rows	$r-1$	M_R	$\sigma^2 + \dfrac{\sum_i n_{i.}\alpha_i^2}{r-1}$
Columns	$c-1$	M_C	$\sigma^2 + \dfrac{\sum_j n_{.j}\beta_j^2}{c-1}$
Interaction	$(r-1)(c-1)$	M_I	$\sigma^2 + \dfrac{\sum_i\sum_j \gamma_{ij}^2}{(r-1)(c-1)}$
Error	$n-rc$	M_E	σ^2
Total	$n-1$		

In this case, the last two expressions on the right hand side are the interaction and error sums of squares, S_I and S_E, with $(r-1)(c-1)$ and $(n - rc)$ d.f. respectively. The expected values of the mean squares M_I and M_E are presented in Table 3.12. A test for the interaction is provided by the F-ratio M_I/M_E.

3.10 Random effects and proportional frequencies

For the case in which α_i, β_j and γ_{ij} of the model in (3.1) are randomly distributed, expected values of the mean squares obtained from (3.44) are presented in Table 3.13. In these expressions,

$$A = \frac{n}{r-1}(1 - \sum_i \frac{n_{i.}^2}{n^2}), \qquad B = \frac{n}{c-1}(1 - \sum_j \frac{n_{.j}^2}{n^2}) ,$$

$$C = A \sum_j n_{.j}^2/n^2, \qquad D = B \sum_i n_{i.}^2/n^2 ,$$

and

$$G = \frac{n}{(r-1)(c-1)}[\sum_i \frac{n_{i.}}{n}(1 - \frac{n_{i.}}{n})][\sum_j \frac{n_{.j}}{n}(1 - \frac{n_{.j}}{n})] . \qquad (3.45)$$

The residual mean square M_E is unbiased for σ^2. Unbiased estimators for the remaining variance components are

$$\hat{\sigma}_\alpha^2 = [GM_R - CM_I + (C - G)M_E]/AG , \qquad (3.46)$$

$$\hat{\sigma}_\beta^2 = [GM_C - DM_I + (D - G)M_E]/BG , \qquad (3.47)$$

and

$$\hat{\sigma}_\gamma^2 = (M_I - M_E)/G . \qquad (3.48)$$

Wilk and Kempthorne (1955) present the expected mean squares of the type given in Table 3.13 for the case of α_i, β_j and γ_{ij} fixed or random. H.F. Smith (1951) discusses the tests of hypotheses for the mixed model with proportional frequencies.

3.11 Further applications

1. *Penicillin manufacturing*
An experiment to assess the variability among samples of penicillin manufactured by the B. Subtilis method was described in

Table 3.13. *Expected mean squares for the case of proportional frequencies; random effects*

Source	d.f.	MS	EMS
Rows	$r-1$	M_R	$A\sigma_\alpha^2 + C\sigma_\gamma^2 + \sigma^2$
Columns	$c-1$	M_C	$B\sigma_\beta^2 + D\sigma_\gamma^2 + \sigma^2$
Interaction	$(r-1)(c-1)$	M_I	$G\sigma_\gamma^2 + \sigma^2$
Residual	$n-rc$	M_E	σ^2
Total	$n-1$		

Davies (1967). Twenty four plates with six sample solutions on each were placed in the incubator and the concentration of penicillin in the solutions was measured at the end of the experiment. The two-way model with both factors random and with one observation in each cell was used for the analysis.

2. *White-cell counts*

Chamberlain and Turner (1952) described a study to investigate the errors occurring in observing the number of white cells per unit volume of blood. The two-way classification with two pipettes, two hemacytometers and four observations in each cell was employed. Both the factors were considered to be random.

3. *Textile production*

Daniels (1938) described one of the earliest applications of the two-way random effects model for examining different methods used for processing wool.

4. *Inter-laboratory studies*

Davies (1967) contains the description of experiments conducted to assess the differences among laboratories. In one experiment, three measurements were made on the densities of two types of chalk in each of eleven laboratories. The two-way random effects model

with interaction between the types of chalk and the laboratories
was used for the analysis.

Results of several experiments of the above type have been described in the literature. Crowder (1992), for instance, examined the suitability of the two-way random effects model with fixed or random effects for analyzing the measurements made on a tensile property on four bars of an alloy by each of six laboratories.

5. *Preparation of dyes*

In another experiment described by Davies (1967), strength or coloring power of six batches of a dye was assessed in two laboratories. The two-way random effects model with interaction was used for analyzing the data from the experiment.

6. *Genetic applications*

For some types of genetic applications, Fry (1992) discusses the relevance of the assumptions for the mixed model.

Exercises

3.1. From the data in Table 3.2, find 95 percent confidence limits for σ_α^2/σ^2 when the rows representing the sites are considered to be random and the columns representing the trace metals are fixed.

3.2. Average the two observations in each cell of Table 3.2. For the case of the rows random and columns fixed, estimate σ_α^2 and find the S.E. of the estimate. Compare these figures with the estimate and S.E. obtained in Section 3.3.

3.3. Find 95 percent confidence limits for σ_α^2/σ^2 from the averaged observations in Exercise 3.2 and compare them with the limits obtained in Exercise 3.1.

3.4. (a) For the test for σ_α^2 described in Section 3.3.1, find the power when $\sigma_\alpha^2/\sigma^2 = 1$, 2 and 10. (b) Does this power increase with the number of rows, columns or the observations in the cells ?

3.5. Average the observations in the cells of Table 3.4. (a) Estimate σ_α^2 and σ_β^2 and (b) find the S.E.s of the estimators. Compare the results with the estimates and S.E.s found in Example 3.2.

3.6. An inter-laboratory assay on calcium in blood serum is de-

Table 3.14. *Calcium in blood serum*

Laboratories	2.0A	2.0B	2.5	3.0A	3.0B
			Concentrations		
A	68	72	333	579	579
	71	68	321	576	571
	67	64	326	578	579
	66	67	317	572	567
D	122	122	386	656	658
	117	110	381	663	658
	119	111	381	659	661
	120	116	387	665	660
M	68	53	292	573	577
	48	52	314	560	574
	42	54	301	552	568
	52	52	310	561	549

Source: Brown, Healy and Kearns (1981)

scribed in Example 3.3. Readings for three of the laboratories are presented in Table 3.14; 1000 was subtracted from each reading. With the model in (3.2), estimate the variance components and find the sample S.E.'s when (a) the laboratories are random and (b) both the laboratories and the target values are random.

3.7. Consider the balanced model in (3.1) with interaction and the data in Table 3.4. Test for the interaction when the row and column effects are fixed.

3.8. Average the two observations in the cells of Table 3.4 and perform Tukey's test for additivity.

3.9. Consider the balanced model with interaction and the data in Table 3.4. For the case of both the rows and columns random, estimate σ_γ^2 and find the sample S.E. of the estimator.

3.10. (a) For the mixed effects model with interaction described in Section 3.7, find an expression for the covariance of β_j and γ_{ij}. (b) Assume that the variances of γ_{ij} are the same and the covariances of γ_{ij} and $\gamma_{i'j}(i \neq i')$ are the same for the rows, and show that the correlation between γ_{ij} and $\gamma_{i'j}$ is given by $-1/(r-1)$.

3.11. For the data in Table 3.9, estimate the column effects from (3.32) and find the adjusted column SS. Note that this SS should be the same as that found in Example 3.4.

3.12. Consider the unbalanced design with interaction described in Section 3.8.4. The adjusted row and column SS for the data in Table 3.9 are presented in Examples 3.3 and 3.4. With these SS, test for the significance of the interaction.

3.13. Derive (a) the expressions for the expected mean squares in Table 3.13 and (b) the estimators for the variance components in (3.46)–(3.48).

3.14. (a) Derive expressions for the variances of the estimators in (3.46)–(3.48). (b) Describe the procedures for finding approximations to the distributions of these estimators. (c) Describe the procedures for testing the hypotheses regarding the variance components from these estimators.

CHAPTER 4

Designs of experiments

4.1 Introduction

A number of interesting designs have been developed for agricultural, genetic, industrial, medical, and other types of research. The differences among a set of treatments and also the classification factors of these designs are examined through models of the type presented in the last two chapters.

Randomized blocks and **Latin squares**, which are particular types of the two-way and three-way classifications, are widely used in practice. When the number of treatments in an experiment exceeds the number of units in a block, the treatments can be examined through the balanced incomplete block design (BIBD). In this chapter, inferences for these designs with fixed or random effects are presented. The designs with hierarchical or nested classifications will be presented in Chapter 5. Further extensions and modifications of the designs presented in these two chapters can be found in the references provided in Section 1.4.

4.2 Randomized blocks

The differences among the effects of a set of diets on weights can be examined by administering the diets randomly to individuals in different weight classes. In this experiment, the individuals are the experimental units, the diets are the treatments and the weight classes form the blocks. The effects of the treatments are examined after eliminating from them the differences among the blocks. To examine the differences among the treatments effectively, the units in a block should be close to each other but differ from the units in the remaining blocks. In agricultural experiments on fertilizer treatments or varieties of food grains, fields with different soil fertilities are used as blocks. Schools in different geographical regions or classes in the schools can be used as blocks in an experiment on teaching methods. To test the effectiveness of a set of medical treatments, hospitals in a region or groups of volunteers can take the place of blocks.

The model for the above type of experiments with b blocks and t treatments is

$$y_{ij} = \mu + \beta_i + \tau_j + \varepsilon_{ij}, \tag{4.1}$$

$i = 1, 2, \ldots, b$ and $j = 1, 2, \ldots t$. The block and treatment effects are denoted by β_i and τ_j respectively and it is assumed that $\sum_i \beta_i = 0$ and $\sum_j \tau_j = 0$. The residual ε_{ij} is assumed to have mean zero and variance σ^2.

The above model is the same as the two-way classification model in (3.2) with $m = 1$ observation in each cell. The rows and columns, for instance, correspond to the blocks and treatments. The difference is that, in randomized block experiments, treatments are randomized in blocks and the residual is formed by the interaction of the blocks and treatments.

Denoting the block, treatment and overall means by $\bar{y}_{i.}$, $\bar{y}_{.j}$ and $\bar{y}$ respectively, the total SS can be expressed as

$$\sum_i \sum_j (y_{ij} - \bar{y})^2 = t \sum_i (\bar{y}_{i.} - \bar{y})^2 + b \sum_j (\bar{y}_{.j} - \bar{y})^2$$
$$+ \sum_i \sum_j (y_{ij} - \bar{y}_{i.} - \bar{y}_{.j} + \bar{y})^2 . \tag{4.2}$$

The expressions on the right hand side are the between blocks, between treatments and residual SS respectively: S_B, S_T and S_E with $(b-1)$, $(t-1)$ and $(b-1)(t-1)$ d.f. respectively. Expectations of the means squares M_B, M_T and M_E for the case of fixed rows and columns are presented in the fourth column of Table 4.1. The test for the equality of the block effects is given by M_B/M_E, which follows the F-distribution with $(b-1)$ and $(b-1)(t-1)$ d.f. Similarly, the test for the equality of the treatments is given by M_T/M_E, which follows the F-distribution with $(t-1)$ and $(b-1)(t-1)$ d.f.

In some applications, the blocks are considered to be a random sample from a large population. In this case, β_i is assumed to have

Table 4.1. *Expected values of the mean squares*

Source	d.f.	MS	Fixed	Random
Blocks	$b-1$	M_B	$\sigma^2 + \dfrac{t\sum_i \beta_i^2}{(b-1)}$	$\sigma^2 + t\sigma_b^2$
Treatments	$t-1$	M_T	$\sigma^2 + \dfrac{b\sum_j \tau_j^2}{(t-1)}$	$\sigma^2 + b\sigma_t^2$
Residual	$(b-1)(t-1)$	M_E	σ^2	σ^2
Total	$bt-1$			

zero mean and variance σ_b^2. An unbiased estimator of this variance is $\hat{\sigma}_b^2 = (M_B - M_E)/t$. Similarly, if the treatments are random, τ_j is assumed to have zero mean and variance σ_t^2. An unbiased estimator of this variance is $\hat{\sigma}_t^2 = (M_T - M_E)/b$.

Example 4.1 Blood pressure treatments

Persons with systolic and diastolic blood pressures in the ranges 140–159/90–99, 160–179/100–109, 180–209/110–119, and 210/120 mmHg are considered to be having the mild, moderate, severe and very severe type of hypertension. The initial measurements and final figures obtained after administering randomly four treatments (t_1, t_2, t_3, t_4) to four persons of the first three types of hypertension are presented in Table 4.2. The first treatment represents a placebo which serves the purpose of a control. The normal levels of 120/80 mmHg are subtracted from the figures.

For this hypothetical example, as presented in Table 4.2, the systolic pressure in the three blocks after administering the treatments are (20, 10, 5, 4), (10, 15, 6, 28) and (16, 45, 20, 18) respectively. The block, treatment and grand totals respectively are (39, 59, 99), (93, 35, 26, 43) and 197. For this data, the different SS obtained from (4.2) and the tests for the hypotheses that $\beta_i = 0$ for $i = 1, 2, \ldots, b$ and $\tau_j = 0$ for $j = 1, 2, \ldots, t$ appear in Table 4.3. The differences among the blocks and also the treatments are highly significant.

Table 4.2. *Randomized blocks and blood pressure treatments*

Mild	t_1	t_4	t_2	t_3
	28, 15	35, 15	30, 10	30, 15
	20, 8	10, 0	5, 2	4, 6
Moderate	t_2	t_4	t_3	t_1
	45, 25	58, 25	50, 25	45, 20
	10, 5	15, 10	6, 4	28, 15
Severe	t_3	t_1	t_2	t_4
	57, 40	55, 30	80, 25	75, 35
	16, 10	45, 28	20, 15	18, 10

The two figures in each cell are the systolic and diastolic blood pressures; the normal levels (120, 80) are subtracted from the figures. The two lines in each block refer to the initial and final measurements.

Table 4.3. *ANOVA for the hypertension treatments*

Source	d.f.	SS	MS	F
Blocks	2	466.67	233.34	15.33
Treatments	3	898.92	299.64	19.69
Residual	6	91.33	15.22	
Total	11	1456.92		

4.3 Balanced incomplete block design (BIBD)

The randomized block experiments cannot be used if the number of treatments t is larger than the number of units or plots k in each block. Even when $t = k$, the units within each block may not be close to each other if the number of treatments is large. In such situations, this design will not be effective for testing the treatment differences.

The balanced incomplete block design (BIBD) is frequently used when $k < t$. In this design, each treatment is replicated r times and each pair of treatments appears in the same number λ of the blocks. The following two conditions are satisfied for t, b, k, r and λ.

(1) Since each treatment is replicated r times, and there are k plots in each block, $tr = bk$.

(2) Each treatment has $(t-1)$ pairs and it appears with the pairs $\lambda(t-1)$ times. A specified treatment has $(k-1)$ pairs in the block in which it appears, and each one of them is replicated r times. Hence, $\lambda(t-1) = r(k-1)$.

Notice that $r \geq k$ and $b \geq t$. This design can be constructed with $t!/k!(t-k)!$ blocks, each of them containing a different combination of the treatments. However, the conditions for this design can sometimes be satisfied with a smaller number of blocks. For instance, if an experiment consists of $t = 6$ treatments and each block contains $k = 3$ plots or experimental units, the design can be constructed with 20 blocks. In this case, $r = bk/t = 10$ and $\lambda = (k-1)r/(t-1) = 4$.

As an alternative, $\lambda = 2r/5$, and it will be equal to 2, an integer, if $r = 5$. Now, $b = tr/k = 10$. This BIBD requires only half as many blocks as the above design.

Fisher and Yates (1953), Cochran and Cox (1957), and Davies (1967) present tables for the BIBDs. The second reference contains illustrations of this design for agricultural experiments. The last reference describes one of the earliest applications of the BIBD for industrial experimentation. In this application, four chemical compounds for manufacturing tires were tested and the four tires of an automobile with $k = 3$ treads on each formed the blocks. In this case, $r = 4$ and $\lambda = 2$. A design of this type with $b = t$ is also known as the symmetric BIBD.

4.3.1 Estimation of the block and treatment effects

The model in (4.1) represents this design. For the least squares method, we minimize

$$\phi = \sum_{ij} n_{ij}(y_{ij} - \mu - \beta_i - \tau_j)^2 , \qquad (4.3)$$

where $n_{ij} = 1$ if the jth treatment appears in the ith block, and zero otherwise. Note that the number of treatments appearing in the ith block is $n_{i.} = \sum_j n_{ij} = k$, and the number of blocks containing the jth treatment is $n_{.j} = \sum_i n_{ij} = r$. Denote the block, treatment and overall totals by $y_{i.}$, $y_{.j}$ and $y_{..}$ respectively.

From the minimization of (4.3), the estimating equations are

$$tr\hat{\mu} + k\sum_i \hat{\beta}_i + r\sum_j \hat{\tau}_j = y_{..} \,, \tag{4.4}$$

$$k\hat{\mu} + k\hat{\beta}_i + \sum_j n_{ij}\hat{\tau}_j = y_{i.} \tag{4.5}$$

and

$$r\hat{\mu} + \sum_i n_{ij}\hat{\beta}_i + r\hat{\tau}_j = y_{.j} \,. \tag{4.6}$$

These $b+t+1$ equations resemble (3.25), (3.26) and (3.27) for the unbalanced two-way cross classification. Information on the block and treatment effects is contained in all three equations. Notice that sum of of the second set of equations over the blocks or the third set over the treatments is equal to the first equation. Thus, only $(b + t - 1)$ of these equations are linearly independent. The estimators for the block and treatment effects can be obtained with the constraints $\sum_i \beta_i = 0$ and $\sum_j \tau_j = 0$.

4.3.2 Treatments adjusted for blocks

Noting that $n_{ij}^2 = n_{ij}$, $\sum_i n_{ij}n_{il} = \lambda$ for $j \neq l$ and $\sum_j \hat{\tau}_j = 0$, from (4.5),

$$r\hat{\mu} + \sum_i n_{ij}\hat{\beta}_i + (r - \lambda)\hat{\tau}_j/k = \left(\sum_i n_{ij}y_{i.}\right)/k \,. \tag{4.7}$$

Subtracting this equation from (4.6),

$$(\lambda t/k)\hat{\tau}_j = y_{.j} - \left(\sum_i n_{ij}y_{i.}\right)/k = Q_j \,. \tag{4.8}$$

The right hand side expression is the **adjusted** treatment total. Note that $\sum_j Q_j = 0$ and these t equations are linearly dependent. Note also that $\sum_i n_{ij}y_{i.}$ is the sum of the totals of the blocks in which the jth treatment appears.

With the estimators obtained from (4.4)–(4.6), the residual SS is

$$S_E = \sum_{ij} n_{ij}(y_{ij} - \hat{\mu} - \hat{\beta}_i - \hat{\tau}_j)^2$$

$$= (\sum_{ij} y_{ij}^2 - \frac{y_{..}^2}{bk}) - (\frac{\sum_i y_{i.}^2}{k} - \frac{y_{..}^2}{bk}) - \sum_j \hat{\tau}_j Q_j . \qquad (4.9)$$

The second expression is obtained by substituting for $(\hat{\mu} + \hat{\beta}_i)$ from (4.5) in the first expression. The three terms on the right hand side respectively are the total SS, unadjusted block SS, $S_B(unadj.)$, and the adjusted treatment SS, $S_T(adj.)$. With the assumption of normality for ε_{ij}, S_E/σ^2 has a χ^2-distribution with $(tr - b - t + 1)$ d.f.

When $\tau_j = 0$ for $j = 1, 2, \ldots, t$, the least squares estimators are obtained from (4.3) with the constraint $\sum_i \hat{\beta}_i = 0$ are $\hat{\mu} = y_{..}/bk$ and $(\hat{\mu} + \hat{\beta}_i) = y_{i.}/k$. The residual SS now becomes

$$S_H = (\sum_{ij} y_{ij}^2 - \frac{y_{..}^2}{bk}) - (\frac{\sum_i y_{i.}^2}{k} - \frac{y_{..}^2}{bk}) . \qquad (4.10)$$

Subtracting (4.10) from (4.9), the adjusted treatment SS is

$$S_T(adj.) = \sum_j \hat{\tau}_j Q_j = (k/\lambda t) \sum_j Q_j^2 . \qquad (4.11)$$

We note that independent of S_E/σ^2, $S_T(adj.)/\sigma^2$ follows a χ^2-distribution with $(t - 1)$ d.f. The expected values of the mean squares $M_T = S_T(adj.)/(t - 1)$ and $M_E = S_E/(tr - b - t + 1)$ are presented in Table 4.4. The ratio $F = M_T/M_E$ follows the F-distribution with $(t - 1)$ and $(tr - b - t + 1)$ d.f., and it provides a test for the above hypothesis on the treatments.

Example 4.2 Automobile mileages

In an experiment on automobile mileages, four test drivers traveled 1000 miles on each of four brands of automobile (A, B, C, D). The speeds varied from 25 to 60 miles per hour. Averages of the mileages are presented in Table 4.5. For this hypothetical example of the BIBD, the automobile brands are the treatments and the test drivers are the blocks. As can be seen, $t = 4$, $r = 3$, $b = 4$ and $k = 3$.

Table 4.4. *Expected mean squares*

Source	d.f.	SS	EMS
S_B(unadj.)	$b-1$	$\dfrac{\sum_i y_{i.}^2}{k} - \dfrac{y_{...}^2}{bk}$	
S_T(adj.)	$t-1$	$\sum_j \hat{\tau}_j Q_j$	$\sigma^2 + [\frac{\lambda t}{k(t-1)}]\sum_j \tau_j^2$
S_E	$tr-b-t+1$	By subtraction	σ^2
Total	$tr-1$	$\sum_i \sum_j y_{ij}^2 - \dfrac{y_{...}^2}{bk}$	

The totals for the treatments A, B, C and D respectively are 72, 68, 100 and 76, with a grand total of G $= 316$. From the above table, the adjusted treatment totals for A, B, C and D respectively are

$$Q_1 = 72 - (80 + 88 + 66)/3 = -6 \ ,$$

$$Q_2 = 68 - (88 + 66 + 82)/3 = -32/3 \ ,$$

$$Q_3 = 100 - (80 + 88 + 82)/3 = 50/3$$

and

$$Q_4 = 76 - (80 + 66 + 82)/3 = 0.$$

Multiplying each of these totals by $(k/\lambda t) = 3/8$, the adjusted treatment effects are $\hat{\tau}_1 = -2.25$, $\hat{\tau}_2 = -4$, $\hat{\tau}_3 = 6.25$ and $\hat{\tau}_4 = 0$.

The adjusted treatment SS is $(3/8)(3848/9) = 160.33$. The remaining sums of squares and the F-ratio for testing that $\hat{\tau}_j = 0$ for the four treatments is presented in Table 4.6. The differences among the treatment effects are highly significant.

4.3.3 Variances of the estimators

From (4.8),

$$V(Q_j) = (r + \frac{kr}{k^2} - 2\frac{r}{k})\sigma^2 = \frac{(k-1)r}{k}\sigma^2 \ . \tag{4.12}$$

Table 4.5. *Gasoline mileages*

| | | Automobile brands | | | | Block totals |
		A	B	C	D	
	1	22	-	32	26	80
Test	2	28	26	34	-	88
Driver	3	22	20	-	24	66
	4	-	22	34	26	82

Table 4.6. *ANOVA for the automobile brands*

Source	d.f.	SS	MS	F
Blocks(unadj.)	3	86.67	28.89	
Treatments(adj.)	3	160.33	53.44	34.9
Residual	5	7.67	1.53	
Total	11	254.67		

Since $\sum_j Q_j = 0$,

$$V(\sum_j Q_j) = tV(Q_j) + t(t-1)Cov(Q_j, Q_l) = 0 \qquad (4.13)$$

for $(j \neq l)$, and hence $Cov(Q_j, Q_l) = -V(Q_j)/(t-1)$. Thus,

$$V(Q_j - Q_l) = 2\frac{t}{t-1}V(Q_j) = 2\frac{\lambda t}{k}\sigma^2 \qquad (4.14)$$

and

$$V(\hat{\tau}_j - \hat{\tau}_l) = \left(\frac{k}{\lambda t}\right)^2 V(Q_j - Q_l) = 2\frac{k}{\lambda t}\sigma^2 = 2\frac{\sigma^2}{rE}, \qquad (4.15)$$

where $E = t(k-1)/k(t-1)$ is known as the efficiency factor of the BIBD. This variance will be small compared, for instance, to the randomized block design (RB), if $\sigma^2 < E\sigma_{RB}^2$.

4.4 Blocks adjusted for treatments

From (4.5) and (4.6),

$$
\begin{aligned}
k\beta_i \;-\; &\frac{1}{r}\sum_j \left(n_{ij} \sum_i n_{ij}\beta_i \right) \\
=\; &\frac{t(r-1)}{b-1}\beta_i \\
=\; &y_{i.} - \frac{1}{r}\sum_j n_{ij}y_{.j} = B_i \ ,
\end{aligned}
\tag{4.16}
$$

where B_i are the **adjusted** block totals. Note that $\sum_i B_i = 0$ and these totals are linearly dependent. From (4.16), unbiased estimators of the block effects are

$$
\hat{\beta}_i = \frac{b-1}{t(r-1)}B_i \ .
\tag{4.17}
$$

Substituting for $(\hat{\mu}+\hat{\tau}_j)$ from (4.6) in the first expression of (4.9), the residual SS becomes

$$
S_E = \left(\sum_{ij} y_{ij}^2 - \frac{y_{..}^2}{bk}\right) - \left(\frac{\sum_j y_{.j}^2}{r} - \frac{y_{..}^2}{bk}\right) - \sum_i \hat{\beta}_i B_i \ .
\tag{4.18}
$$

If $\beta_i = 0$ for $i = 1, 2, \ldots, b$, the least squares estimators obtained from (4.3) with $\sum_j \hat{\tau}_j = 0$ are $\hat{\mu} = y_{..}/bk$ and $(\hat{\mu}+\hat{\tau}_j) = y_{.j}/r$. The residual SS now is

$$
S_H = \left(\sum_{ij} y_{ij}^2 - \frac{y_{..}^2}{bk}\right) - \left(\frac{\sum_j y_{.j}^2}{r} - \frac{y_{..}^2}{bk}\right) \ .
\tag{4.19}
$$

Note that the second term is the unadjusted treatments SS, $S_T(unadj.)$. Subtracting (4.19) from (4.18), the adjusted block SS is

$$
S_B(adj.) = \sum_i \hat{\beta}_i B_i = \frac{b-1}{t(r-1)}\sum_i B_i^2 \ .
\tag{4.20}
$$

Table 4.7. *Expected mean squares*

Source	d.f.	SS	Fixed
$S_B(\text{adj})$	$b-1$	$\sum_i \hat{\beta}_i B_i$	$\sigma^2 + \frac{t(r-1)}{(b-1)^2}\sum_i \beta_i^2$
$S_T(\text{unadj})$	$t-1$	$\sum_j y_{\cdot j}^2/k - y_{\cdot\cdot}^2/bk$	
S_E	$tr-b-t+1$	From Table 4.4	σ^2
Total	$tr-1$		

As seen in (3.38) for the unbalanced two-way classification,

$$S_T(adj.) + S_B(unadj.) = S_B(adj.) + S_T(unadj.) . \qquad (4.21)$$

The expected value of the mean square $M_B(adj.) = S_B(adj.)/(b-1)$ is presented in the fourth column of Table 4.7. The ratio $F = M_B/M_E$, which follows the F-distribution with $(b-1)$ and $(tr - b - t + 1)$ d.f., provides a test for the above hypothesis regarding the block effects.

Example 4.3 Automobile mileages

For the gasoline mileages, from the block totals in Table 4.5 and from (4.16), the adjusted block totals are $B_1 = -8/3, B_2 = 8, B_3 = -6$ and $B_4 = 2/3$. From (4.17), the estimates for the block effects are $b_1 = -1, b_2 = 3, b_3 = -9/4$ and $b_4 = 1/4$. The adjusted blocks SS from (4.20) is 40.33. As given in Table 4.8, the F-ratio for testing $\beta_i = 0$ for $i = 1, 2, \ldots, 4$ is 8.78. This figure indicates that the gasoline mileages can also be different for the drivers. The speeds at which they drive and their driving habits can be two reasons for this difference.

Table 4.8. *ANOVA for blocks*

Source	d.f.	SS	MS	F
Blocks(adj.)	3	40.33	13.44	8.78
Treatments(unadj.)	3	206.67		
Residual	5	7.67	1.53	
Total	11	254.67		

4.5 Interblock estimators for the BIBD

For some applications, the blocks are selected randomly. For instance, the drivers in Example 4.2 can be considered to be selected randomly. In this case β_i is assumed to have a normal distribution with mean zero and variance σ_β^2. The expected value of $M_B(adj.)$ for this case is $\sigma^2 + t(r-1)\sigma_\beta^2/(b-1)$. An unbiased estimator of σ_β^2 is

$$\hat{\sigma}_\beta^2 = \frac{b-1}{t(r-1)}[M_B(adj) - M_E] . \tag{4.22}$$

For the gasoline mileage example, with the block and residual mean squares in Table 4.8, $\hat{\sigma}_\beta^2 = (3/8)(13.44 - 1.53) = 4.47$. From the F-ratio of 8.78, it can be inferred that σ_β^2 is significantly larger than zero.

4.5.1 Interblock estimators of treatment effects

The **intrablock** estimators for τ_j are given by (4.8). When the blocks are random, as Yates (1940) demonstrated, alternative estimators for τ_j can be obtained from the block totals. From (4.1),

$$y_{i.} = k\mu + k\beta_i + \sum_j n_{ij}\tau_j + \varepsilon_{i.} , \tag{4.23}$$

where $\varepsilon_{i.} = \sum_j \varepsilon_{ij}$. Note that

$$V(y_{i.}) = k^2\sigma_\beta^2 + k\sigma^2 = k(k\sigma_\beta^2 + \sigma^2) . \tag{4.24}$$

Minimizing

$$\phi = \sum_i (y_i. - k\mu - \sum_j n_{ij}\tau_j)^2 \,, \tag{4.25}$$

the estimators for μ and τ_j are given by

$$bk\tilde{\mu} + r \sum_j \tilde{\tau}_j = y.. \tag{4.26}$$

and

$$kr\tilde{\mu} + \sum_i [n_{ij} \sum_j n_{ij}\tilde{\tau}_j] = \sum_i n_{ij}y_i.. \tag{4.27}$$

With $\sum_j \tilde{\tau}_j = 0$, these equations become

$$\tilde{\mu} = y../bk = \overline{y} \tag{4.28}$$

and

$$kr\tilde{\mu} + (r - \lambda)\tilde{\tau}_j = \sum_i n_{ij}y_i.. \tag{4.29}$$

Thus,

$$\tilde{\tau}_j = \frac{\sum_i n_{ij}y_i. - kr\overline{y}}{r - \lambda} = \frac{Q'_j}{r - \lambda}\,. \tag{4.30}$$

This **interblock** estimator is unbiased for τ_j.

The interblock estimator for $(\tau_j - \tau_l)$ for $(j \neq l)$ is

$$\tilde{\tau}_j - \tilde{\tau}_l = \frac{\sum_i (n_{ij} - n_{il})y_i.}{r - \lambda}\,, \tag{4.31}$$

and its variance is

$$\begin{aligned}
V(\tilde{\tau}_j - \tilde{\tau}_l) &= \frac{1}{(r - \lambda)^2} \sum_i (n_{ij}^2 + n_{il}^2 - 2n_{ij}n_{il})V(y_i.) \\
&= \frac{2(r - \lambda)V(y_i.)}{(r - \lambda)^2} = \frac{2k(k\sigma_\beta^2 + \sigma^2)}{r - \lambda}\,. \tag{4.32}
\end{aligned}$$

An unbiased estimator of this variance is obtained by replacing σ^2 by the error mean square M_E and σ_β^2 by its estimator in (4.22).

4.5.2 Combining intra and inter block estimates

From the expression for Q_j in (4.8) and Q'_j in (4.30),

$$Cov(Q_j, Q'_j) = Cov(y._j, \sum_i n_{ij}y_i.) - Cov(y._j, \sum_j y._j)$$

$$- \frac{1}{k}V\left(\sum_i n_{ij}y_{i.}\right) + \frac{1}{k}V\left(\sum_i n_{ij}y_{i.}\right) . \qquad (4.33)$$

The first two covariances are equal to $r\sigma^2$. Hence the covariance of (Q_j, Q'_j) or $(\hat{\tau}_j, \tilde{\tau}_j)$ vanishes.

A combined estimator for $(\tau_j - \tau_l)$ which is unbiased is

$$(\tau_j^* - \tau_l^*) = \omega(\hat{\tau}_j - \hat{\tau}_l) + (1 - \omega)(\tilde{\tau}_j - \tilde{\tau}_l) , \qquad (4.34)$$

where ω is a constant. Denoting the variances in (4.15) and (4.32) by V_1 and V_2,

$$V(\tau_j^* - \tau_l^*) = \omega^2 V_1 + (1 - \omega)^2 V_2 . \qquad (4.35)$$

The minimum of this variance is obtained when $\omega = V_2/(V_1 + V_2)$. For this optimum estimator, with this weight,

$$V_{opt}(\tau_j^* - \tau_l^*) = \frac{V_1 V_2}{V_1 + V_2} = \frac{1}{(1/V_1) + (1/V_2)} . \qquad (4.36)$$

For this estimator and its variance, σ^2 is obtained from the residual mean square M_E, and σ_β^2 from (4.22).

4.6 Latin squares

As seen in Chapter 3, in the two-way cross-classification, differences among the rows or columns are examined after eliminating the other factor and in the randomized blocks experiments, block differences are eliminated before examining the differences of the treatments. In a Latin square arrangement, which is a three-way classification, treatment differences are examined after eliminating the differences of the remaining two factors. The model for this design is

$$y_{ijk} = \mu + \alpha_i + \beta_j + \gamma_k + \varepsilon_{ijk} \qquad (4.37)$$

where α_i, β_j and γ_k for $(i, j, k) = 1, 2, \ldots, t$ represent the rows, columns and treatments. The error ε_{ijk} is assumed to follow the normal distribution with mean zero and variance σ^2.

One of the earliest industrial applications of the Latin squares, described in Davies (1967), was related to a wear-testing experiment of four materials. The positions and runs for a machine were represented by the columns and rows. Tippett (1931) describes a Latin square experiment for testing the differences among four treatments for warp breakage. Four looms and four time periods were used for the columns and rows. All the three factors were considered to be random.

Denoting the row, column, treatment and overall means by $\overline{y}_{i..}$, $\overline{y}_{.j.}$, $\overline{y}_{..k}$, and $\overline{y}$, the total sum of squares can be expressed as

$$
\sum_{ijk}(y_{ijk} - \overline{y})^2 = t\sum_i(\overline{y}_{i..} - \overline{y})^2 + t\sum_j(\overline{y}_{.j.} - \overline{y})^2
$$

$$
+ \ t\sum_j(\overline{y}_{.j.} - \overline{y})^2
$$

$$
+ \ \sum_{ijk}(y_{ijk} - \overline{y}_{i..} - \overline{y}_{.j.} + 2\overline{y})^2 . \tag{4.38}
$$

The right hand side expressions are the row, column, treatment and residual sums of squares, S_R, S_C, S_T and S_E respectively.

Each of the first three sums of squares has $(t-1)$ d.f. and the expectations of their mean squares M_R, M_C and M_T are $\sigma^2 + t\sum_i \alpha_i^2/(t-1), \sigma^2 + t\sum_j \beta_j^2/(t-1)$ and $\sigma^2 + t\sum_k \gamma_k^2/(t-1)$ respectively. The residual sum of squares has $(t-1)(t-2)$ d.f. and its mean square $M_E = S_E/(t-1)(t-2)$ is unbiased for σ^2. Tests of hypotheses for the three factors are provided by the ratios of their mean squares to the residual mean square.

Example 4.4 Automobile mileages

The mileages for four brands of automobile A, B, C and D, driven by four test drivers at four speed levels is presented in Table 4.9. The analysis of this data is presented in Table 4.10. Since $F(3,6; 0.05) = 4.76$, the differences among the automobile brands are significant at the 5 percent level. The differences among the drivers or speeds are not significant.

4.6.1 Random effects

If all the three factors of a Latin square are considered to be random, they are assumed to have zero means and variances σ_α^2, σ_β^2 and σ_γ^2 respectively. The expectations of the three mean squares M_R, M_C and M_T in this case are $\sigma^2 + t\sigma_\alpha^2, \sigma^2 + t\sigma_\beta^2$ and $\sigma^2 + t\sigma_\gamma^2$. Hence, unbiased estimators of σ_α^2, σ_β^2 and σ_γ^2 are $(M_R - M_E)/t$, $(M_C - M_E)/t$ and $(M_T - M_E)/t$ respectively.

Table 4.9. *Latin square for automobile mileages*

		Speed: Miles per hour				Row totals
		<30	30–40	40–60	>60	
	1	C32	D26	A22	B22	102
Test driver	2	A28	B20	C30	D30	108
	3	D24	A22	B26	C34	106
	4	B22	C34	D26	A24	106
Column totals		106	102	104	110	
Treatment totals		A96	B90	C130	D106	

Table 4.10. *ANOVA for the automobile mileages*

Source	d.f.	SS	MS	F
Drivers	3	4.75	1.58	
Speeds	3	8.75	2.92	
Brands	3	232.75	77.58	7.82
Residual	6	59.50	9.92	
Total	15	305.75		

In some applications, only some of the three factors are considered to be random. If the automobile brands of Example 4.4 are considered to be random, from Table 4.10, $\hat{\sigma}_\gamma^2 = (77.58 - 9.92)/4 = 16.92$.

Note that if one of the columns of Table 4.9 is deleted, the resulting design becomes a BIBD with rows as blocks. Similarly, if one of the rows is deleted, it becomes a BIBD with columns as blocks.

4.6.2 Replications of a Latin square

Consider an experiment to compare t medical treatments, for instance, two headache remedies. This experiment can be conducted through a Latin square with the rows representing the times for administering the treatments, for instance, morning and afternoon, and the columns representing the patients or volunteers selected for the experiment. This experiment can be replicated R times, for instance, for six days or for four pairs of volunteers.

For this replicated Latin square, the Total, Treatment and Replication SS have $(Rt^2 - 1)$, $(t - 1)$ and $(R - 1)$ d.f. respectively. Since there are Rt rows and Rt columns, the row SS and column SS have $(Rt - 1)$ d.f. each. The SS due to rows in replications with $R(t - 1)$ d.f. is obtained by subtracting the replication SS from the row SS. Similarly, the SS due to columns in replications with $R(t - 1)$ d.f. is obtained by subtracting the Replication SS from the column SS. The Residual SS has $(t - 1)(Rt - R - 1)$ d.f. For the experiment on headache remedies, $t = 2$ and $R = 6$, and hence the Residual SS has 5 d.f.

The **cross-over** design can be considered as an alternative to a replicated Latin square. In these designs, which are frequently used for medical research, the treatments are administered to the experimental units in a sequence. For instance, one of the headache remedies mentioned above can be administered in the morning and the second one in the afternoon to six patients. The order of administering the treatments is reversed for six more patients. The two rows of this design represent the times of administering the treatments and the columns represent the patients. The Residual SS for this design has 11 d.f., six more than for the Residual SS of the replicated Latin square. For a chronic health disorder, Koch *et al* (1988) examine two medical treatments and a placebo through a cross-over design.

4.6.3 Orthogonal Latin squares

In an orthogonal or Graeco Latin square, treatment differences can be examined after eliminating factors in addition to the rows and columns. For instance, in the automobile mileage example, four types of road conditions can be included as an additional factor.

This design can be formed by superimposing, on the original design, a Latin square with Greek letters. Each Greek letter appears once in each row and each column and once with each Roman letter. The sums of squares due to rows, columns, Roman letters, Greek letters and the residual are all independently distributed. The d.f. for the residual for this design will be $(t-1)(t-3)$. A maximum of $(t-1)$ orthogonal Latin squares of size t can be constructed, and replications of this design are needed to have sufficient number of d.f. for the Residual SS.

Exercises

4.1 (a) From the data in Table 4.2, test the differences of the treatments for lowering the diastolic blood pressures. (b) Estimate σ_t^2 assuming that the treatments are selected randomly from a large population, and find the S.E. of the estimate.

4.2 From the data in Table 4.2, test the differences among the treatments for the percentages of the declines in the (a) systolic pressures and (b) diastolic pressures.

4.3 From the data of the BIBD experiment in Table 4.5, find the intrablock estimate for the difference of the effects of the mileages for automobile brands A and B and estimate its S.E.

4.4 From the data in Table 4.5, find the interblock estimate for the difference of the effects of the mileages for A and B and estimate its S.E.

4.5 Find the combined intra and inter block estimate for the difference in the mileages for A and B and estimate its S.E.

4.6 Form a BIBD by deleting the first column of the Latin square in Table 4.9. For the automobile brands, find the adjusted totals and their effects. Find the S.E. for the difference of the effects.

4.7 Consider the mileage data in Table 4.9 to be the observations from a randomized block experiment with the rows as blocks. (a) Test for the significance of the mileages for the four brands of automobiles. (b) Consider the blocks to be random and estimate σ_β^2.

4.8 For a second replication of the Latin square in Table 4.9, the mileages for the four drivers respectively are (34, 28, 20, 24), (32, 24, 32, 34), (28, 22, 24, 32) and (26, 34, 26, 24); the positions of the four brands remain the same. Describe the model for this replicated Latin square and test for the differences of the brands, speeds and drivers.

4.9 Following the procedure in Section 4.5.2, find the combined intra and interblock estimator for the contrast $\sum_j c_j \tau_j$, where $\sum_j c_j = 0$, and find its variance.

CHAPTER 5

Nested classifications

5.1 Introduction

Experimental designs with hierarchical classifications are frequently used in agricultural, genetic, industrial, medical and other types of research. Cochran's (1939) description of a sampling scheme for estimating wheat production provides an illustration. In this procedure, samples of farms were selected from six districts. At the next stage, samples of fields were selected from each of the selected farms. At the final stage, measurements on the yield of wheat were obtained from sample 'paths' in each of the selected fields.

For demographic, political and socioeconomic studies, samples of geographic regions, counties, districts and towns are selected in a succession. Similar procedures of sampling are employed in geographical studies on rock formation, mineral deposition and soil erosion. These types of designs are also used in studies related to water and atmospheric pollution, and also in environmental and ecological studies.

Nested designs can be balanced or unbalanced, the classification factors can be fixed or random, and the variances of the random effects can be equal or unequal. Some interesting nested designs from the literature are presented in Section 5.6.

5.2 Three-stage nested design

The one-way random effects model described in Chapter 2 represents a two-stage design. To illustrate a three-stage nested design, we selected three regions in the U.S., two states from each of the three regions and four counties from each of the chosen states. The

Table 5.1. *Number of physicians per 10,000 persons*

	Region 1		Region 2		Region 3	
	MA	NY	CA	OR	AR	TX
Counties	27	34	17	9	8	9
	18	11	25	19	11	10
	13	16	14	11	6	8
	15	24	23	18	9	10
State totals	73	85	79	57	34	37
Region totals	158		136		71	

Source: American Hospital Association Guide to Health Care; 1989 figures. The six states are Massachusetts, New York, California, Oregon, Arkansas and Texas.

number of physicians in the sample counties are presented in Table 5.1. The regions, states and counties can be represented by A, B and C respectively. This type of design is also known as the two factor design; A and B are the two factors.

5.2.1 Model for the three-stage design

The three stages of the above illustration can be represented by the model

$$y_{ijk} = \mu + \alpha_i + \beta_{(i)j} + \epsilon_{(ij)k}, \tag{5.1}$$

$i = 1, 2, \ldots, a; j = 1, 2, \ldots, b;$ and $k = 1, 2, \ldots, c$. The parentheses for the subscripts indicate the nesting of the classifications. For the above illustration, $a = 3$ regions, $b = 2$ states in each region and $c = 4$ counties per state. The residual is usually denoted by ε_{ijk} without the parentheses.

The means of the samples at the second and third stages are $\bar{y}_{ij.} = \sum_k y_{ijk}/c$ and $\bar{y}_{i..} = \sum_{jk} y_{ijk}/bc = \sum_j \bar{y}_{ij.}/b$. The overall mean of the $n = abc$ sample observations is

$$\bar{y} = \sum_{ijk} y_{ijk}/n = \sum_{ij} \bar{y}_{ij.}/ab = \sum_i \bar{y}_{i..}/a. \tag{5.2}$$

The Total SS can be expressed as

$$\sum_{ijk}(y_{ijk} - \bar{y})^2$$
$$= \sum_{ijk}\left[(\bar{y}_{i..} - \bar{y}) + (\bar{y}_{ij.} - \bar{y}_{i..}) + (y_{ijk} - \bar{y}_{ij.})\right]^2$$
$$= bc\sum_i(\bar{y}_{i..} - \bar{y})^2 + c\sum_{ij}(\bar{y}_{ij.} - \bar{y}_{i..})^2$$
$$+ \sum_{ijk}(\bar{y}_{ijk} - \bar{y}_{ij.})^2$$
$$= S_A + S_B + S_E . \tag{5.3}$$

The first term on the right hand side S_A is the SS among the first-stage units, and S_B is the SS among the second-stage units nested in the first-stage units. The last term S_E is the Residual SS.

The Total SS can be obtained from $\sum_{ijk} y_{ijk}^2 - G^2/n$ where $G = \sum_{ijk} y_{ijk}$ is the total of all the n observations and G^2/n is the correction factor (C.F.). Denoting the first stage totals by $T_{i..} = \sum_{jk} y_{ijk}$, $S_A = \sum_i T_{i..}^2/bc - C.F.$ Similarly, the SS among the second stage units is $\sum_{ij} T_{ij.}^2/c - C.F.$, where $T_{ij.} = \sum_k y_{ijk}$. Now, S_B is obtained by subtracting S_A from this expression. Finally, S_E is obtained by subtracting S_A and S_B from the Total SS. The mean squares of these three SS are $M_A = S_A/(a-1)$, $M_B = S_B/a(b-1)$ and $M_E = S_E/ab(c-1)$, and M_E is unbiased for σ^2.

5.2.2 Fixed effects

If A is considered to be nonrandom, it is assumed that $\sum_i \alpha_i = 0$. In this case,

$$E(M_A) = bc\sum_i \alpha_i^2/(a-1) + \sigma^2 . \tag{5.4}$$

Table 5.2. *Sums of squares and mean squares for the physicians data*

Source	d.f.	SS	MS
Regions	2	511.58	255.79
States in regions	3	79.63	26.54
States	5	591.21	118.24
Residual	18	586.75	32.6
Total	23	1177.96	

With the assumption of normality for ε_{ijk}, a test for $\alpha_i = 0, i = 1, 2, \ldots, a$ is provided by $F = M_A/M_E$, which follows the F-distribution with $(a-1)$ and $ab(c-1)$ d.f.

Similarly, if B is considered to be nonrandom, it is assumed that $\sum_j \beta_{(i)j} =$ for $i = 1, 2, \ldots, a$. In this case,

$$E(M_B) = c \sum_i \sum_j \beta_{(i)j}^2 / a(b-1) + \sigma^2 . \tag{5.5}$$

A test for $\beta_{(i)j} = 0$ is given by $F = M_B/M_E$, which follows the F-distribution with $a(b-1)$ and $ab(c-1)$ d.f.

Example 5.1 Physicians in the states

The mean squares for the data of Table 5.1 are presented in Table 5.2. To test $\alpha_i = 0$, $F = 255.79/32.6 = 7.85$, which is significant at the 0.01 level. From the relative magnitudes of the mean squares for the states in regions and the residual, we find that the states in the regions do not differ significantly.

5.2.3 Random effects

If both A and B are considered to be random, α_i and $\beta_{(i)j}$ are assumed to have independent normal distributions with zero means and variances σ_α^2 and σ_β^2 respectively. They are also assumed to be independent of ε_{ijk}. The expectations of the mean squares for this case are presented in Table 5.3.

A test for $\sigma_\alpha^2 = 0$ is given by $F = M_A/M_B$, which follows the

Table 5.3. *Expectations of the mean squares*

Source	d.f.	SS	MS	EMS
A	$a-1$	S_A	M_A	$bc\sigma_\alpha^2 + c\sigma_\beta^2 + \sigma^2$
B in A	$a(b-1)$	S_B	M_B	$c\sigma_\beta^2 + \sigma^2$
C in B	$ab(c-1)$	S_E	M_E	σ^2
Total	$n-1$			

F-distribution with $(a-1)$ and $a(b-1)$ d.f. Similarly, a test for $\sigma_\beta^2 = 0$ is given by $F = M_B/M_E$, which follows the F-distribution with $a(b-1)4$ and $ab(c-1)$ d.f.

From Table 5.2, we find that for testing $\sigma_\alpha^2 = 0$, $F = 255.79/26.54 = 9.64$. Since $F(2,3; 0.05) = 9.55$, we find that σ_α^2 is significantly larger than zero. As can be seen from the relative magnitudes of M_B and M_E, σ_β^2 is not significantly larger than zero.

With $F = M_B/M_E$, confidence limits for σ_β^2/σ^2 can be obtained from $(F-F_u)/cF_u$ and $(F-F_l)/cF_l$, where F_l and F_u are the lower and upper percentage points of the F-distribution with $a(b-1)$ and $ab(c-1)$ d.f. for the specified probability.

5.2.4 Mixed effects

If A is fixed but B is random, it is assumed that $\sum_i \alpha_i = 0$ and the assumption for $\beta_{(i)j}$ remains the same as in the above section. In this case, $E(M_A) = bc \sum_i \alpha_i^2/(a-1) + c\sigma_\beta^2 + \sigma^2$, and the expectations of M_B and M_E remain the same as in Table 5.3. The test for $\alpha = 0$ is given by $F = M_A/M_B$.

5.3 Estimation of variance components

If both A and B are random as described in Section 5.2.3, an unbiased estimator of σ_α^2 is

$$\hat{\sigma}_\alpha^2 = (M_A - M_B)/bc \ , \tag{5.6}$$

which has variance

$$V(\hat{\sigma}_\alpha^2) = \frac{2}{(bc)^2} \left[\frac{E^2(M_A)}{a-1} + \frac{E^2(M_B)}{a(b-1)} \right] \ . \tag{5.7}$$

An unbiased estimator of this variance is

$$v(\hat{\sigma}_\alpha^2) = \frac{2}{(bc)^2} \left[\frac{M_A^2}{a+1} + \frac{M_B^2}{a(b-1)+2} \right] \ . \tag{5.8}$$

An unbiased estimator for σ_β^2 is

$$\hat{\sigma}_\beta^2 = (M_B - M_E)/c \ , \tag{5.9}$$

and its variance is

$$V(\hat{\sigma}_\beta^2) = \frac{2}{c^2} \left[\frac{E^2(M_B)}{a(b-1)} + \frac{E^2(M_E)}{ab(c-1)} \right] \ . \tag{5.10}$$

An unbiased estimator of this variance is

$$v(\hat{\sigma}_\beta^2) = \frac{2}{c^2} \left[\frac{M_B^2}{a(b-1)+2} + \frac{M_E^2}{ab(c-1)+2} \right] \ . \tag{5.11}$$

Example 5.2 Physicians in the states

From (5.6) and the mean squares in Table 5.2, $\hat{\sigma}_\alpha^2 = 28.66$. From (5.8) we find that $S.E.(\hat{\sigma}_\alpha^2) = 22.71$. Since the estimate in (5.9) is negative, σ_β^2 may be estimated with a small positive quantity.

5.4 Unbalanced designs

The balanced designs are easy to implement. However, in these designs, the d.f. available for M_A and M_B can be much smaller than for M_E. As can be seen from (5.7) and (5.10), the S.E.s of $\hat{\sigma}_\alpha^2$ and $\hat{\sigma}_\beta^2$ can be large if the d.f. for M_A and M_B are not large.

As an illustration, consider the following three designs with $n = 24$: (1) $a = 4$, $b = 2$ and $c = 3$; (2) $a = 6$, $b = 2$ and $c = 2$; and (3) $a = 8$, $b = 2$, $c_{i1} = 2$ and $c_{i2} = 1$. The first two designs are balanced. In the last design, which is unbalanced, samples of sizes two and one are selected from the two second stage units. The d.f. for M_A,

M_B and M_E for these three designs respectively are (1) (3, 4, 16), (2) (5, 6, 12) and (3) (7, 8, 8). The second design with relatively more d.f. for M_A and M_B can be expected to have smaller S.E.s for $\hat{\sigma}_\alpha^2$ and $\hat{\sigma}_\beta^2$ than the first design. The S.E.s of these estimators can be expected to be smaller for the third design than the first two designs. Illustrations of a balanced and two unbalanced designs are presented in Figure 5.1.

A three-stage design can be unbalanced at both the second and the third stages. If the numbers of observations at the third stage are unequal, the design in (5.1) with $k = 1, 2, \ldots, c_{ij}$ can represent the observations. The sample means in this case are $\bar{y}_{ij.} = \sum_k y_{ijk}/c_{ij}$, $\bar{y}_{i..} = \sum_{jk} y_{ijk}/c_{i.} = \sum_j c_{ij}\bar{y}_{ij.}/c_{i.}$ and $\bar{y} = \sum_{ijk} y_{ijk}/n$, where $c_{i.} = \sum_j c_{ij}$.

The Total SS can now be expressed as

$$
\begin{aligned}
\sum_{ijk}(y_{ijk} - \bar{y})^2 &= \sum_i c_{i.}(\bar{y}_{i..} - \bar{y})^2 + \sum_{ij} c_{ij}(\bar{y}_{ij.} - \bar{y}_{i..})^2 \\
&+ \sum_{ijk}(y_{ijk} - \bar{y}_{ij.})^2 \\
&= Q + \sum_i Q_i + \sum_{ij} Q_{ij} \\
&= S_A + S_B + S_E, \qquad\qquad (5.12)
\end{aligned}
$$

where $Q = \sum_i c_{i.}(\bar{y}_{i..} - \bar{y})^2$, $Q_i = \sum_j c_{ij}(\bar{y}_{ij.} - \bar{y}_{i..})^2$ and $Q_{ij} = \sum_k(\bar{y}_{ijk} - \bar{y}_{ij.})^2$. The sums of squares $S_A = Q$, $S_B = \sum_i Q_i$ and $S_E = \sum_{ij} Q_{ij}$ as before are the variances among the first stage units, between the second stage units within the first stage units, and between the third stage units within the second stage units respectively. The sum of squares S_A is the same as $\sum_i y_{i..}^2/c_{i.} - C.F.$. Similarly, $\sum_{ij} c_{ij}(\bar{y}_{ij.} - \bar{y}^2) = \sum_{ij} y_{ij.}^2 - C.F.$. The sum of squares S_B is obtained by subtracting S_A from this SS. The mean squares M_A, M_B and M_E are obtained as before by dividing these SS by the corresponding d.f.

5.4.1 Fixed effects

If the first and second stage units are considered to be nonrandom, it is assumed that $\sum_i c_{i.}\alpha_i = 0$ and $\sum_j c_{ij}\beta_{(i)j} = 0$ for

$i = 1, 2, \ldots, a$. In this case,

$$E(M_A) = \sum_i c_i. \alpha_i^2/(a-1) + \sigma^2 \; , \qquad\qquad (5.13)$$

and

$$E(M_B) = \sum_i \sum_j c_{ij} \beta_{(i)j}^2/a(b-1) + \sigma^2 \; . \qquad\qquad (5.14)$$

An unbiased estimator for σ^2 is again provided by M_E. The ratios M_A/M_E and M_B/M_E provide the tests for $\alpha_i = 0$ and $\beta_{(i)j} = 0$ respectively.

5.4.2 Random effects

If the units at the first and second stages are random, the assumptions for α_i and $\beta_{(i)j}$ remain the same as described in Section 5.2.3. Now,

$$
\begin{aligned}
E(S_A) \;=\; & (n - \sum_i c_{i.}^2/n)\sigma_\alpha^2 \\
& + \sum_i \sum_j \left[(c_{ij}^2/c_{i.}) - c_{ij}^2/n \right] \sigma_\beta^2 + (a-1)\sigma^2
\end{aligned}
\qquad (5.15)
$$

and

$$E(S_B) = (n - \sum_i \sum_j c_{ij}^2/c_{i.})\sigma_\beta^2 + a(b-1)\sigma^2 \; . \qquad (5.16)$$

As before M_E is an unbiased estimator for σ^2. Unbiased estimators for σ_α^2 and σ_β^2 are obtained from these equations by removing the expectation signs on the left hand sides.

Goldsmith and Gaylor (1970) compare the above type of estimators for different amounts of unbalancedness at the second and third stages. Heckler (1993) suggests estimators that utilize prior information on the variance components and iterative procedures. Mahamunulu (1963), Blischke (1966) and Leone *et al.* (1968) examine the variances of the estimators of the variance components of the three-stage nested designs.

Example 5.3 Assay of the enzyme lypase

For the sake of illustrating the unbalanced design, we consider the data from Heckler and Rao (1985) for assaying lypase. This en-

Figure 5.1. *Designs with different amounts of balancing for a total sample of size 12*

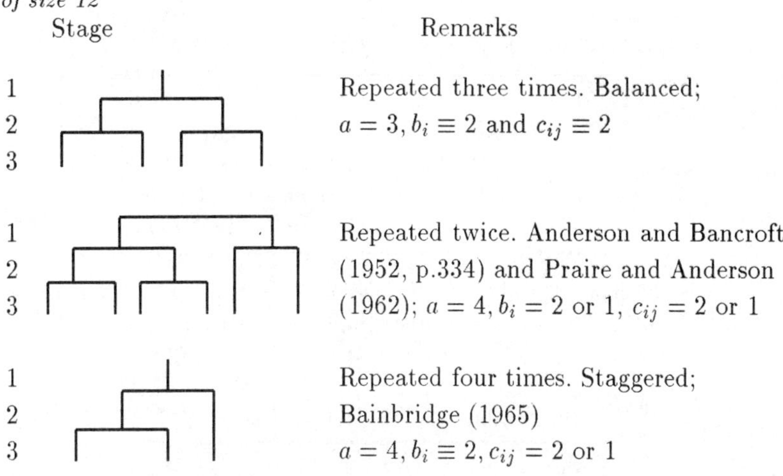

zyme is used for certain types of medical diagnosis, and the different stages of its preparation can affect the required specifications. Three laboratories preparing the enzyme were randomly selected for the experiment. Four weeks were randomly assigned for each of the laboratories. Measurements were obtained in the mornings and evenings of the sample days chosen from the selected weeks. In Table 5.4, only the averages for the days are presented; 450 is subtracted from each of the averages.

For this unbalanced design, $a = 3, b = 4, c_{ij} = 3$ for $j = 1$ and $c_{ij} = 2$ for $j = 2, 3, 4$. The ANOVA for this data is presented in Table 5.5. As can be seen, $\hat{\sigma}^2 = 60.72$. Since $c_{i.} = 9, \sum_i c_{i.}^2 = 243$, $\sum_i \sum_j c_{ij}^2 = 63$ and $\sum_i \sum_j (c_{ij}^2/c_{i.}) = 7$, from (5.15) and (5.16), $\hat{\sigma}_\alpha^2 = 138.94$ and $\hat{\sigma}_\beta^2 = 53.95$.

5.5 Unequal variances and sample sizes

For this general case, the variances of $\beta_{(i)j}$ are unequal and they can be denoted by σ_i^2. The variance of ε_{ijk}, denoted by σ_{ij}^2 can also be unequal. Equality of σ_i^2 and of σ_{ij}^2 are usually assumed. In

Table 5.4. Lypase measurements

Week	Laboratory 1	Laboratory 2	Laboratory 3
1	43.4, 46.2, 46.5	7.0, 7.8, 15.7	22.4, 15.5 29.7
2	37.0, 16.6	32.4, 16.8	25.4, 23.1
3	23.6, 33.6	13.4, 9.6	22.9, 0.6
4	51.0, 52.4	23.9, 19.3	18.4, 3.7

Week totals: $(136.1, 30.5, 67.6)$; $(53.6, 49.2, 48.5)$; $(57.2, 23.0, 23.5)$; $(103.4, 43.2, 22.1)$. Lab Totals: $350.3, 145.9, 161.7$

Table 5.5. ANOVA for lypase

Source	d.f.	SS	MS
Labs	2	2874.03	1437.02
Weeks in labs	9	1625.45	180.61
Weeks	11	4499.48	409.04
Residual	15	910.82	60.72
Total	26	5410.30	

some practical situations, these assumptions may not be valid.

The model in (5.1) with $i = 1, 2, \ldots, a, j = 1, 2, \ldots, b$ and $k = 1, 2, \ldots, c_{ij}$ represents this general situation. The variances of α_i, $\beta_{(i)j}$ and ε_{ijk} are $\sigma_\alpha^2, \sigma_i^2$ and σ_{ij}^2 respectively.

The Total SS can be expressed as in (5.12). The expectations of the different quadratic forms are as follows.

$$E(Q_{ij}) = (c_{ij} - 1)\sigma_{ij}^2 \ ,$$

$$E(Q_i) = (c_{i.} - \sum_j c_{ij}^2/c_{i.})\sigma_i^2 + \sum_j (1 - c_{ij}/c_{i.})\sigma_{ij}^2 \ ,$$

$$E(Q) = (n - \sum_i \frac{c_{i.}^2}{n})\sigma_\alpha^2 + \sum_i \sum_j \frac{(n - c_{i.})c_{ij}(c_{ij}\sigma_i^2 + \sigma_{ij}^2)}{nc_{i.}} \ . \tag{5.17}$$

Unbiased estimators for all the variance components can be obtained by removing the expectation signs and solving the resulting equations. If some of the σ_{ij}^2 or σ_i^2 are equal, the corresponding quadratic forms Q_{ij} or Q_i are added up before finding the estimators.

5.6 Further illustrations

A number of applications of the nested designs appeared in the literature, some of which are described below.

1. Fabric differences

Tippett (1931) presents an experiment for examining the properties of four fabrics. Three tests were performed on each of the fabrics, and each test was repeated four times. The mean squares for the fabrics, tests in fabrics, and residual were 17.54, 30.53 and 7.54 respectively.

2. Blood pressure measurements

The Hypertension Prevention Trial was conducted in four U.S. clinics during 1983-1986. Canner *et al.* (1991) used data from this study and a nested model to examine errors made in measuring blood pressures of individuals. The effects of the participants in the study, their visits and duplicate measurements on each visit were included in the model and they were all assumed to be random. The analysis showed that the duplicate measurements reduce the standard deviation of the observations by five percent. The authors also found that additional reduction in the variability can be achieved if the duplicate measurements on a participant are made by different persons. The second conclusion, however, is surprising since blood pressure measurements of an individual are usually less variable when they are made by the personal physician or trained persons familiar to the individual.

3. Eye examination

Rosner (1982) uses a nested model to study items such as "intraoccular pressures in persons." The model consists of the effects of groups, individuals in the groups, and the measurements on both the eyes. The groups were considered to be fixed, and the remaining two factors random. For some items, measurement on one of the eyes was missing. For some other items, the condition being examined by the opthalmologist existed in only one of the eyes.

4. Asparagus clones

For patenting asparagus clones, a plant producer used estimates of the means and variances of their important characteristics. For future experimentation, estimates of the variance components of the cladophylls, "the tiny leaves located on the asparagus branches," were also obtained. Trout (1985) describes this interesting application of the variance components model. The study was conducted by selecting in stages, five stalks from a clone, two branches from each stalk, five nodes on each branch, and three cladophylls from each node. Variance components were obtained from the lengths of the 150 cladophylls at the nodes.

5. Textile production

Bainbridge (1965) suggests a staggered design for detecting sources of variation occurring in industrial productions, and illustrates it through a chemical test on a specific textile. From a large number of machines, two were selected on each of forty-two days. The sample from one of the machines was tested by two analysts on different shifts, one of them obtaining duplicate measurements. The sample from the second machine was analyzed only once by an analyst. The data from this experiment were used for studying the variations occurring from (1) changes in the raw material over the days, (2) differences in the machines, (3) long term tests at the different shifts, and (4) short term tests through the duplicate measurements. This four stage design is unbalanced. Praire and Anderson (1962) considered a three-stage design of this type and showed that the MSEs for the first two stages are dependent; see Fig. 5.1.

6. Experimental drugs

Patients with certain diseases are hospitalized and administered suitable medical treatments. Experimental drugs can be examined by administering them to the patients receiving each of the treatments. In a split-plot experiment, the treatments and drugs are considered to be the main and sub-plot treatments respectively.

7. Spectral density

Jackson and Lawton (1969) examined the consequences of estimating a spectral density through a nested classification.

8. Animal breeding

In several experiments on animal breeding, each of a sample of sires are randomly mated to samples of dames. The observations from the offspring are analyzed through the model for a nested design.

Exercises

5.1 For samples of four counties from each of the six states described in Section 5.2, the number of hospital beds per 10,000 persons are presented in Table 5.6. Since the figure of 141 for the third county in New York is relatively large, replace it by the average 45 of the remaining counties. Test for differences among the regions and among the states. Estimate the variance components for both these factors and find the S.E.s of the estimates.

5.2 As seen in Example 5.1, the differences among the states in the regions are negligible. Consider the one-way random effects model for the states and test whether the corresponding variance component is larger than zero.

5.3 (a) Describe the model for the first illustration of Section 5.6 on fabrics. (b) With the mean squares given by Tippett, estimate the variance components. (c) Examine the hypothesis that the variance component for the tests on fabrics is significantly larger than zero.

5.4 (a) Describe the model for the asparagus study described in the fourth illustration of Section 5.6. and find the expected values of the relevant MSEs. (b) From the data on the 150 cladophylls, Trout found the MSEs for the stalks, branches in stalks, nodes in branches and the residual to be 101.86, 21.11, 5.23 and 3.35 respectively. Estimate the variance components of the first three factors and test whether they are significantly larger than zero.

5.5 (a) Describe the model for the staggered design described in the fifth illustration of Section 5.6.

5.6 The data in Table 5.7 were presented by Snee (1983) for a four-stage design on a chemical analysis; 90 is subtracted from all the figures. (a) Describe the model for this balanced design. (b) Find the expectations of the SS and the MSEs at the four stages. (c) Estimate the variance components of the operators, samples and

Table 5.6. *Number of hospital beds per 10,000 persons*

	Region 1		Region 2		Region 3	
	MA	NY	CA	OR	AR	TX
	48	55	21	81	20	37
	44	40	30	32	41	49
Counties	60	141	23	45	22	20
	57	40	29	30	22	27

Table 5.7. *Four-stage staggered design*

| | | Run | | | | | |
| | | 1 | | 2 | | 3 | |
Operator	Sample	Test 1	Test 2	Test 1	Test 2	Test 1	Test 2
1	1	66	64	61	64	64	70
	2	58	60	64	67	57	59
2	3	94	94	82	96	91	101
	4	82	86	91	94	85	87

Source: Snee (1983)

tests. (d) Test whether the variance component for the operators is significantly larger than zero.

5.7 (a) Describe the model for the second illustration of Section 5.6 on blood pressures. (b) Find an expression for the average of the blood pressure measurements on an individual and for its variance.

CHAPTER 6

ML and REML estimation

6.1 Introduction

In the ANOVA procedures, as seen in the previous chapters, the variance components are estimated from the different sums of squares. The maximum likelihood (ML) method can also be considered for estimating both the variance components and the fixed parameters. Cochran (1937), for instance, presented this approach for the one-way classification model in (2.1) with unequal residual variances.

The ML procedure for estimating the variance components does not take into account the degrees of freedom lost for estimating the fixed parameters, and the restricted maximum likelihood (REML) method rectifies this situation. For the unbalanced designs, explicit expressions for the ML and REML estimators of the variance components can not be found in general, and iterative procedures have been developed to find these estimators.

6.2 The general mixed model

The models in the previous chapters with the fixed, random or mixed effects in general can be expressed as

$$Y = X\beta + \varepsilon \ . \tag{6.1}$$

In this expression, Y is an $n \times 1$ vector of observations, X is a known $n \times s$ matrix and β is the $s \times 1$ vector of the fixed parameters. The $n \times 1$ vector ε can be expressed as

$$\varepsilon = U_1\xi_1 + U_2\xi_2 + \cdots + U_p\xi_p = U\xi \ , \tag{6.2}$$

where the $n \times n_i$ matrices U_i, $i = 1, 2, \ldots, p$ are known and the $n_i \times 1$ vectors ξ_i represent the random effects and residuals, with

$n = \sum_i n_i$. Note that $U = (U_1 \vdots U_2 \vdots \cdots \vdots U_p)$, $\xi' = (\xi'_1 \vdots \xi'_2 \vdots \cdots \vdots \xi'_p)$, and $\varepsilon = U\xi$.

It is assumed that ξ_i follows a normal distribution with $\mathrm{E}(\xi_i) = 0$, $\mathrm{E}(\xi_i \xi'_i) = \sigma_i^2 I_i$, and $\mathrm{E}(\xi_i \xi'_j) = 0$ for $i \neq j$. The variance-covariance or dispersion matrix of Y is

$$\Sigma = E(\varepsilon\varepsilon') = E(U\xi)(U\xi)' = \sum_1^p \sigma_i^2 V_i \ , \tag{6.3}$$

where $V_i = U_i U'_i$. When ξ_i, $i = 1, 2, \ldots, p$ has a normal distribution, Y follows the multinormal distribution with mean $X\beta$ and dispersion Σ.

The model in (6.1) with $U_p = I$ and $\mathrm{E}(\xi_p \xi'_p) = \sigma_p^2 I$ was considered by Hartley and J.N.K. Rao (1967) for describing the ML procedure for the ANOVA models. It can also represent a number of other practical situations such as the regression models with random coefficients.

For the models with variance and covariance components,

$$\varepsilon = U_1 \xi_1 + \cdots + U_p \xi_p + U_{p+1} \xi_{p+1} + \cdots + U_{p+q} \xi_{p+q} \ . \tag{6.4}$$

The assumptions on ξ_i for $i = 1, 2, \ldots, p$ are the same as above. In addition, it is assumed that $\mathrm{E}(\xi_i) = 0$, $\mathrm{E}(\xi_i \xi'_i) = \Sigma_i$ for $i = p+1, p+2, \ldots, p+q$, and ξ_i and ξ_j for $(i \neq j) = 1, 2, \ldots, p+q$ are uncorrelated. From (6.4) the dispersion of Y now becomes

$$\Sigma = \sum_1^p \sigma_i^2 V_i + \sum_{p+1}^{p+q} U_i \Sigma_i U'_i \ . \tag{6.5}$$

For an alternative formulation of (6.2), ξ_i for $i = 1, 2, \ldots, p-1$ representing the variance components are assumed to have multinormal distributions with zero means and dispersions $\sigma_i^2 S_i$. The residual vector $U_p \xi_p$ is assumed to have a multinormal distribution with zero mean and dispersion $\sigma_p^2 E$. With these assumptions, Y has a multinormal distribution with mean $X\beta$ and dispersion $\Sigma = \sum_1^{p-1} \sigma_i^2 U_i S_i U'_i + \sigma_p^2 E = USU' + \sigma_p^2 E$, where S is diagonal with $\sigma_i^2 S_i$. Harville (1977), for instance, considered this type of representation for the mixed models.

6.3 Maximum likelihood estimators (MLEs)

If $\boldsymbol{Y}$ has a multinormal distribution with mean $\boldsymbol{X\beta}$ and covariance $\boldsymbol{\Sigma}$, the likelihood function is given by

$$L = L(\boldsymbol{\beta}, \boldsymbol{\sigma}; \boldsymbol{Y}) = \frac{e^{-[(\boldsymbol{Y} - \boldsymbol{X\beta})'\boldsymbol{W}(\boldsymbol{Y} - \boldsymbol{X\beta})]/2}}{(2\pi)^{n/2}|\boldsymbol{\Sigma}|^{1/2}}, \qquad (6.6)$$

where $\boldsymbol{\sigma}$ is the vector of the p variances and $\boldsymbol{W} = \boldsymbol{\Sigma}^{-1}$.
The natural logarithm $l = \log_e L$ is

$$l = -\frac{n}{2}ln(2\pi) - \frac{1}{2}ln(|\boldsymbol{\Sigma}|) - \frac{1}{2}(\boldsymbol{Y} - \boldsymbol{X\beta})'\boldsymbol{W}(\boldsymbol{Y} - \boldsymbol{X\beta}) . \quad (6.7)$$

For the mixed model represented by (6.1)-(6.3), the first partial derivatives of (6.7) with respect to $\boldsymbol{\beta}$ and σ_i^2 are

$$\frac{\partial l}{\partial \boldsymbol{\beta}} = \boldsymbol{X}'\boldsymbol{W}\boldsymbol{Y} - \boldsymbol{X}'\boldsymbol{W}\boldsymbol{X\beta} = \boldsymbol{X}'\boldsymbol{W}(\boldsymbol{Y} - \boldsymbol{X\beta}) \qquad (6.8)$$

and

$$\frac{\partial l}{\partial \sigma_i^2} = -\frac{1}{2}tr(\boldsymbol{W}\boldsymbol{V}_i) + \frac{1}{2}(\boldsymbol{Y} - \boldsymbol{X\beta})'\boldsymbol{W}\boldsymbol{V}_i\boldsymbol{W}(\boldsymbol{Y} - \boldsymbol{X\beta}) . \quad (6.9)$$

Setting these derivatives to zero,

$$\widehat{\boldsymbol{\beta}} = (\boldsymbol{X}'\boldsymbol{W}\boldsymbol{X})^- \boldsymbol{X}'\boldsymbol{W}\boldsymbol{Y} \qquad (6.10)$$

and

$$\begin{aligned}
tr(\boldsymbol{W}\boldsymbol{V}_i) &= (\boldsymbol{Y} - \boldsymbol{X\beta})'\boldsymbol{W}\boldsymbol{V}_i\boldsymbol{W}(\boldsymbol{Y} - \boldsymbol{X\beta}) \\
&= \boldsymbol{Y}'\boldsymbol{R}\boldsymbol{V}_i\boldsymbol{R}\boldsymbol{Y} = \boldsymbol{e}'\boldsymbol{V}_i\boldsymbol{e} = g_i ,
\end{aligned} \qquad (6.11)$$

In this expression $\boldsymbol{e} = \boldsymbol{R}\boldsymbol{Y}$, where $\boldsymbol{Q}_W = \boldsymbol{I} - \boldsymbol{X}(\boldsymbol{X}'\boldsymbol{W}\boldsymbol{X})^- \boldsymbol{X}'\boldsymbol{W}$, and $\boldsymbol{R} = \boldsymbol{W}\boldsymbol{Q}_W$. The MLEs are obtained from these two equations with the constraints that the variance components be nonnegative. Iterative procedures for finding the estimates through this procedure are presented in Sections 6.6–6.8.

6.3.1 The information matrix

Denote the elements of $\boldsymbol{\beta}$ or $\boldsymbol{\sigma}$ by θ_i, and the vector of all the parameters by $\boldsymbol{\theta}$. For the MLE of θ_i, the second partial derivative $\partial^2 l/\partial\theta_i^2$ should be negative. Similarly, for the MLE of the entire vector $\boldsymbol{\theta}$, the matrix $D(\boldsymbol{\theta})$ of the second partial derivatives $d_{ij} = \partial^2 l/\partial\theta_i\partial\theta_j$ should be negative definite.

The information in an estimator of θ_i is given by

$$\mathcal{I}(\theta_i) = -E(\partial^2 l/\partial\theta_i^2) = E(\partial l/\partial\theta_i)^2 \ . \tag{6.12}$$

The Cramer–Rao lower bound to the variance of an estimator of θ_i is given by $1/\mathcal{I}(\theta_i)$. For the case of more than one parameter, the elements of the information matrix $I(\theta)$ are given by $-E(d_{ij}) = E[(\partial l/\partial\theta_i)(\partial l/\partial\theta_j)]$.

From (6.8),

$$\partial^2 l/\partial\beta^2 = -\boldsymbol{X'WX} \ , \tag{6.13}$$

and hence $I(\beta) = \boldsymbol{X'WX}$. From (6.9),

$$\begin{aligned}
\partial^2 l/\partial\sigma_i^2\partial\sigma_j^2 \ &= \ (1/2)tr(\boldsymbol{WV_iWV_j}) \\
&\quad - \ (\boldsymbol{Y}-\boldsymbol{X\beta})'\boldsymbol{WV_iWV_jW}(\boldsymbol{Y}-\boldsymbol{X\beta}) \ .
\end{aligned} \tag{6.14}$$

From the expectation of (6.14), we find that $I(\boldsymbol{\sigma})$ consists of the elements $(1/2)tr\ \boldsymbol{WV_iWV_j}$, that is, it is given by the matrix $(1/2)(tr\ \boldsymbol{WV_iWV_j})$. From (6.9), we note that $(\partial/\partial\beta)(\partial l/\partial\sigma_i^2) = -\boldsymbol{X'WV_iW}(\boldsymbol{Y}-\boldsymbol{X\beta})$, and its expectation vanishes. Thus,

$$I(\boldsymbol{\theta}) = \begin{pmatrix} I(\boldsymbol{\beta}) & \bullet \\ \bullet & I(\boldsymbol{\sigma}) \end{pmatrix} = \begin{pmatrix} \boldsymbol{X'WX} & \bullet \\ \bullet & (1/2)\boldsymbol{H} \end{pmatrix} \ , \tag{6.15}$$

where $\boldsymbol{H}$ is the $p \times p$ matrix with elements $tr(\boldsymbol{WV_iWV_j})$.

The above derivations can be obtained from the general procedures described in C.R. Rao (1973; p.52 and Section 5a).

6.3.2 MLEs for the one-way classification

For the one-way classification represented by (2.1), let the $n_i \times 1$ column vector $\boldsymbol{Y_i}$ denote the n_i observations of the ith group. With the assumption of normality for y_{ij}, the likelihood is

$$L_i = L(\mu, \sigma_i^2, \sigma_\alpha^2; \boldsymbol{Y_i}) = \frac{e^{-\frac{1}{2}(\boldsymbol{Y_i}-\mu\boldsymbol{1})'\boldsymbol{W_i}(\boldsymbol{Y_i}-\mu\boldsymbol{1})}}{(2\pi)^{n_i/2}|\boldsymbol{\Sigma_i}|^{1/2}} \ , \tag{6.16}$$

where $\boldsymbol{1}$ denotes the $n_i \times 1$ vector of unities. The diagonal and off-diagonal elements of the $n_i \times n_i$ covariance matrix $\boldsymbol{\Sigma_i}$ are equal to $(\sigma_\alpha^2 + \sigma_i^2)$ and σ_α^2 respectively, and $\boldsymbol{W_i} = \boldsymbol{\Sigma_i}^{-1}$.

The likelihood for the $n = \sum_i n_i$ observations of the k groups is

$$L \;=\; \prod_i L_i \;=$$

$$\frac{exp[-\frac{1}{2}\sum_i\sum_j \frac{(y_{ij}-\mu)^2}{\sigma_i^2} + \frac{\sigma_\alpha^2}{2}\sum_i \frac{n_i(\overline{y}_i-\mu)^2}{\sigma_i^2 v_i}]}{\prod_i (2\pi)^{n_i/2}|\mathbf{\Sigma}_i|^{1/2}} \;, \qquad (6.17)$$

where $v_i = \sigma_\alpha^2 + \sigma_i^2/n_i$. Noting that $\sum_j (y_{ij}-\mu)^2 = \sum_j (y_{ij}-\overline{y}_i)^2 + n_i(\overline{y}_i - \mu)^2$, and expanding $|\mathbf{\Sigma}_i|$,

$$L = \frac{exp[-\frac{1}{2}\sum_i \frac{(n_i-1)s_i^2}{\sigma_i^2} - \frac{1}{2}\sum_i \frac{(\overline{y}_i-\mu)^2}{v_i}]}{\prod_i [(2\pi)^{n_i/2}(\sigma_i^2)^{\frac{n_i-1}{2}} n_i^{\frac{1}{2}} v_i^{\frac{1}{2}}]} \;. \qquad (6.18)$$

The natural logarithm of L is

$$l \;=\; -\left(\frac{n}{2}\right)ln(2\pi) - \left(\frac{1}{2}\right)\sum_i (n_i-1)ln(\sigma_i^2) - \left(\frac{1}{2}\right)\sum_i ln(n_i v_i)$$

$$-(1/2)\sum_i \frac{(n_i-1)s_i^2}{\sigma_i^2} - (1/2)\sum_i \frac{(\overline{y}_i-\mu)^2}{v_i} \;. \qquad (6.19)$$

The MLEs of μ, σ_i^2 and σ_α^2 are obtained by maximizing (6.18) or (6.19) with respect to these parameters with the constraints that $\sigma_i^2 > 0$ and $\sigma_\alpha^2 \geq 0$. The first partial derivatives of (6.19) are

$$\frac{\partial l}{\partial \mu} \;=\; \sum_i \frac{\overline{y}_i - \mu}{v_i} \;,$$

$$\frac{\partial l}{\partial \sigma_i^2} \;=\; -\frac{n_i-1}{2\sigma_i^2} - \frac{1}{2n_i v_i} + \frac{(n_i-1)s_i^2}{2\sigma_i^4} + \frac{(\overline{y}_i-\mu)^2}{2n_i v_i^2} \;,$$

and

$$\frac{\partial l}{\partial \sigma_\alpha^2} \;=\; -\frac{1}{2}\sum_i \frac{1}{v_i} + \frac{1}{2}\sum_i \frac{(\overline{y}_i-\mu)^2}{v_i^2} \;. \qquad (6.20)$$

If $\sigma_i^2 \equiv \sigma^2$, v_i becomes $(\sigma_\alpha^2 + \sigma^2/n_i)$, and the derivative with respect to σ^2 is given by

$$\frac{\partial l}{\partial \sigma^2} \;=\; -\frac{(n-k)}{2\sigma^2} - \frac{1}{2}\sum_i \frac{1}{n_i v_i} + \frac{\sum_i (n_i-1)s_i^2}{2\sigma^4}$$

$$+ \; \frac{1}{2} \sum_i \frac{(\overline{y}_i - \mu)^2}{n_i v_i^2} \; . \tag{6.21}$$

For the cases of unequal σ_i^2 or unequal n_i, explicit solutions for μ, σ_i^2 and σ_α^2 cannot be obtained from these derivatives. Iterative procedures with the constraints of nonnegativeness for the estimators of the variance components should be used for this purpose.

If $\sigma_i^2 \equiv \sigma^2$ and $n_i \equiv m$, the first derivatives of l are given by

$$\frac{\partial l}{\partial \mu} = \frac{\sum_i(\overline{y}_i - \mu)}{v} \; ,$$

$$\frac{\partial l}{\partial \sigma^2} = -\frac{(n-k)}{2\sigma^2} - \frac{k}{2mv} + \frac{(m-1)\sum_i s_i^2}{2\sigma^4} + \frac{\sum_i(\overline{y}_i - \mu)^2}{2mv^2} \; ,$$

and

$$\frac{\partial l}{\partial \sigma_\alpha^2} = -\frac{k}{2v} + \frac{\sum_i(\overline{y}_i - \mu)^2}{2v^2} \; , \tag{6.22}$$

where $v = \sigma_\alpha^2 + \sigma^2/m$. The solutions from equating these derivatives to zero are

$$\hat{\mu} = \overline{y} = \sum_i \overline{y}_i/k \; ,$$

$$\hat{\sigma}^2 = s^2 = M_W \; ,$$

and

$$\hat{\sigma}_\alpha^2 = \sum_i(\overline{y}_i - \overline{y})^2/k - s^2/m = (k-1)M_B/n - M_W/m \; , \tag{6.23}$$

where M_B and M_W are the between and within mean squares as defined in Chapter 2.

The above solutions for μ and σ^2 are the same as the ANOVA estimators in Section 2.3.1. However, the solution for σ_α^2 in (6.23) is the MLE for σ_α^2 only if it is nonnegative. If it becomes negative, σ_α^2 should be considered to be zero, that is, the random effect α_i should be replaced by a constant, say α. As a result, the MLEs of $\mu' = (\mu + \alpha)$ and σ^2 will now be given by $\overline{y}$ and

$$\hat{\sigma}^2 = \sum_i \sum_j(y_{ij} - \overline{y})^2/n = [(k-1)M_B + (n-k)M_W]/n \; . \tag{6.24}$$

In summary, the MLEs of σ^2 and σ_α^2 are given by the expressions in (6.23) only if $\hat{\sigma}_\alpha^2$ is nonnegative. Otherwise, the MLE of

σ^2 is given by (6.24), and the MLE of σ_α^2 is considered to be zero. Although, the above change of the model was not expressed explicitly, Herbach (1959) showed that these alternative estimators maximize the likelihood. The probability that the solution for σ_α^2 in (6.23) to be nonnegative is

$$P = P[F_{k-1,n-k} \geq k\sigma^2/(k-1)(m\sigma_\alpha^2 + \sigma^2)] \,, \tag{6.25}$$

which increases as σ_α^2 becomes large relative to σ^2.

6.4 Restricted maximum likelihood estimators (REMLs)

In general, the MLEs for the variance components do not take into account the loss in degrees of freedom resulting from the estimation of the fixed effects, and hence they become biased. For instance, consider a linear model represented by (6.1) with the residual vector ε having zero mean and dispersion $\sigma^2\,\boldsymbol{I}$. The MLE for σ^2 is $\hat\sigma^2 = \boldsymbol{Y}'[\boldsymbol{I} - \boldsymbol{X}(\boldsymbol{X}'\boldsymbol{X})^-\boldsymbol{X}']\boldsymbol{Y}/n$ and the bias of this estimator is $-r\sigma^2/n$ where $r(\leq s)$ is the rank of $\boldsymbol{X}$. The underestimation increases with r.

As an alternative, consider an $n \times (n - r)$ matrix $\boldsymbol{A}$ with rank $(n - r)$. The $(n - r)$-vector $\boldsymbol{A}'\boldsymbol{Y}$ with $\boldsymbol{A}'\boldsymbol{X} = \boldsymbol{0}$ is known as the *maximal invariant*. The likelihood with $\boldsymbol{A}'\boldsymbol{Y}$,

$$L_1 = L(\boldsymbol{\sigma}; \boldsymbol{A}'\boldsymbol{Y}) = \frac{e^{-[\boldsymbol{Y}'\boldsymbol{A}(\boldsymbol{A}'\boldsymbol{\Sigma}\boldsymbol{A})^{-1}\boldsymbol{A}'\boldsymbol{Y}]/2}}{(2\pi)^{(n-r)/2}|\boldsymbol{A}'\boldsymbol{\Sigma}\boldsymbol{A}|^{1/2}} \tag{6.26}$$

is known as the **marginal likelihood**, and it does not depend on the fixed parameter $\boldsymbol{\beta}$. Thompson (1962), and Patterson and Thompson (1971) suggested estimating the variance components from this likelihood. Anderson and Bancroft (1952) and Russell and Bradley (1958) considered similar procedures.

As shown by C.R. Rao (1973, p.77), $\boldsymbol{R} = \boldsymbol{A}(\boldsymbol{A}'\boldsymbol{\Sigma}\boldsymbol{A})^{-1}\boldsymbol{A}' = \boldsymbol{W}[\boldsymbol{I} - \boldsymbol{X}(\boldsymbol{X}'\boldsymbol{W}\boldsymbol{X})^-\boldsymbol{X}'\boldsymbol{W}]$, and hence the natural logarithm of L_1 can be expressed as

$$l_1 = -[(n-r)/2]ln(2\pi) - (1/2)ln|\boldsymbol{A}'\boldsymbol{\Sigma}\boldsymbol{A}| - (1/2)\boldsymbol{Y}'\boldsymbol{R}\boldsymbol{Y} \,. \tag{6.27}$$

Note that

$$\frac{\partial ln|\boldsymbol{A}'\boldsymbol{\Sigma}\boldsymbol{A}|}{\partial\sigma_i^2} = tr(\boldsymbol{A}'\boldsymbol{\Sigma}\boldsymbol{A})^{-1}\frac{\partial(\boldsymbol{A}'\boldsymbol{\Sigma}\boldsymbol{A})}{\partial\sigma_i^2}$$

$$= tr(\boldsymbol{A}'\boldsymbol{\Sigma}\boldsymbol{A})^{-1}\boldsymbol{A}'\boldsymbol{V}_i\boldsymbol{A} = tr\boldsymbol{R}\boldsymbol{V}_i \tag{6.28}$$

and

$$\partial(\boldsymbol{Y}'\boldsymbol{R}\boldsymbol{Y})/\partial\sigma_i^2 = -\boldsymbol{Y}'\boldsymbol{R}(\partial\Sigma/\partial\sigma_i^2)\boldsymbol{R}\boldsymbol{Y} = -\boldsymbol{Y}'\boldsymbol{R}\boldsymbol{V}_i\boldsymbol{R}\boldsymbol{Y} \ . \quad (6.29)$$

From these two expressions,

$$\partial l_1/\partial\sigma_i^2 = -(1/2)tr\boldsymbol{R}\boldsymbol{V}_i + (1/2)\boldsymbol{Y}'\boldsymbol{R}\boldsymbol{V}_i\boldsymbol{R}\boldsymbol{Y} \ . \quad\quad\quad (6.30)$$

Setting this expression to zero, the marginal maximum likelihood (MML) or the restricted maximum likelihood (REML) estimators of σ_i^2 are obtained from

$$tr\boldsymbol{R}\boldsymbol{V}_i = \boldsymbol{Y}'\boldsymbol{R}\boldsymbol{V}_i\boldsymbol{R}\boldsymbol{Y} = g_i \ . \quad\quad\quad (6.31)$$

Iterative procedures of the type described in Sections (6.6)–(6.8) with the constraints that $\sigma_i^2 \geq 0$ are used for this purpose.

Notice from (6.31) that the left hand side expression is the expectation of the quadratic form on the right hand side. Secondly, since $\boldsymbol{R}\Sigma\boldsymbol{R} = \boldsymbol{R}$, the left hand side can also be expressed as tr $\boldsymbol{R}\Sigma\boldsymbol{R}\boldsymbol{V}_i$. The estimators for σ_i^2 can now be obtained from

$$\boldsymbol{F}\widehat{\boldsymbol{\sigma}} = \boldsymbol{g} \ , \quad\quad\quad (6.32)$$

where the elements of the $p \times p$ matrix $\boldsymbol{F}$ are given by $tr\boldsymbol{R}\boldsymbol{V}_i\boldsymbol{R}\boldsymbol{V}_j$, and the elements of $\boldsymbol{g}$ by g_i.

From (6.30),

$$\frac{\partial^2 l_1}{\partial\sigma_i^2\partial\sigma_j^2} = (1/2)tr\boldsymbol{R}\boldsymbol{V}_i\boldsymbol{R}\boldsymbol{V}_j - \boldsymbol{Y}'\boldsymbol{R}\boldsymbol{V}_i\boldsymbol{R}\boldsymbol{V}_j\boldsymbol{R}\boldsymbol{Y} \ , \quad\quad (6.33)$$

and

$$\begin{aligned} E\left(\frac{\partial^2 l_1}{\partial\sigma_i^2\partial\sigma_j^2}\right) &= (1/2)tr\boldsymbol{R}\boldsymbol{V}_i\boldsymbol{R}\boldsymbol{V}_j - tr\boldsymbol{R}\boldsymbol{V}_i\boldsymbol{R}\boldsymbol{V}_j\boldsymbol{R}\Sigma \\ &= -(1/2)tr\boldsymbol{R}\boldsymbol{V}_i\boldsymbol{R}\boldsymbol{V}_j \ . \quad\quad (6.34) \end{aligned}$$

The matrix of these expectations is given by $-(1/2)\boldsymbol{F}$.

The design matrix $\boldsymbol{X}$ can be expressed as $(\boldsymbol{X}_1 \vdots \boldsymbol{X}_2)$, where the $(n \times r)$ matrix $\boldsymbol{X}_1$ consists of r linearly independent columns of $\boldsymbol{X}$, and $\boldsymbol{X}_2$ consists of linear combinations of the columns of $\boldsymbol{X}_1$. Thus, $\boldsymbol{A}'\boldsymbol{X} = \boldsymbol{o} \Leftrightarrow \boldsymbol{A}'\boldsymbol{X}_1 = \boldsymbol{o}$, and the MLEs of σ_i^2 can be obtained from maximizing the likelihood $L(\boldsymbol{\sigma}; \boldsymbol{A}'\boldsymbol{Y})$ with $\boldsymbol{A}'\boldsymbol{X}_1 = \boldsymbol{o}$. The estimating equations in this case will be the same as (6.31).

As a further simplification, Harville (1977) mentions that the logarithm of $L(\boldsymbol{\sigma}; \boldsymbol{A'Y})$ with $\boldsymbol{A'X}_1 = \mathbf{o}$ differs only by a constant from

$$l_1^* = -(1/2)ln|\boldsymbol{\Sigma}| - (1/2)ln|\boldsymbol{X}_1'\boldsymbol{W}\boldsymbol{X}_1| - (1/2)\boldsymbol{Y'RY} . \qquad (6.35)$$

The estimating equations (6.31) and (6.32) are also obtained by maximizing this expression.

The fixed parameter β can be estimated from (6.10) by utilizing the estimates of the variance components obtained from (6.32).

6.4.1 REML for the one-way classification

For the model in (2.1), from (6.35),

$$\begin{aligned}
-2l_1^* &= \sum_i (n_i - 1)ln\sigma_i^2 + \sum_i ln(n_i v_i) \\
&\quad + \ln\left[\sum_i (1/v_i)\right] + \sum_i (n_i - 1)s_i^2/\sigma_i^2 \\
&\quad + \sum_i (\bar{y}_i - \bar{y}_w)^2/v_i ,
\end{aligned} \qquad (6.36)$$

where $w_i = 1/v_i$ and $\bar{y}_w = \sum_i w_i\bar{y}_i / \sum_i w_i$. Hence,

$$\begin{aligned}
-2\frac{\partial l_1^*}{\partial \sigma_i^2} &= (n_i - 1)/\sigma_i^2 + (1/n_i v_i) - (1/n_i v_i^2 \sum_i w_i) \\
&\quad -(n_i - 1)s_i^2/\sigma_i^4 - (\bar{y}_i - \bar{y}_w)^2/n_i v_i^2
\end{aligned}$$

and

$$-2\frac{\partial l_1^*}{\partial \sigma_\alpha^2} = \sum_i w_i - \sum_i w_i^2 / \sum_i w_i - \sum_i w_i^2(\bar{y}_i - \bar{y}_w)^2 . \qquad (6.37)$$

Now, the solutions for σ_i^2 and σ_α^2 are obtained from

$$\frac{\sigma_i^2 + (n_i - 1)\sigma_\alpha^2}{v_i\sigma_i^2} - \frac{w_i^2}{n_i \sum_i w_i} = \frac{(n_i - 1)s_i^2}{\sigma_i^4} + \frac{w_i^2(\bar{y}_i - \bar{y}_w)^2}{n_i}$$

and

$$\sum_i w_i - \sum_i w_i^2 / \sum_i w_i = \sum_i w_i^2(\bar{y}_i - \bar{y}_w)^2 . \qquad (6.38)$$

These solutions can also be obtained from (6.31) and (6.32).

If $\sigma_i^2 \equiv \sigma^2$ and $n_i \equiv m$, we find from these solutions that $\hat{\sigma}^2 = s^2$ and

$$\hat{\sigma}_\alpha^2 = (M_B - M_W)/m , \tag{6.39}$$

which are the same as the ANOVA estimators found in Section 2.3.1. If $\hat{\sigma}_\alpha^2$ in (6.39) takes a negative value, as noted by Thompson (1962), it is set to zero and σ^2 is estimated from

$$\hat{\sigma}^2 = [(k-1)M_B + (n-k)M_W]/(n-1) . \tag{6.40}$$

The probability for $\hat{\sigma}_\alpha^2$ in (6.39) to be nonnegative is

$$P = P[F_{k-1,n-k} \geq \sigma^2/(m\sigma_\alpha^2 + \sigma^2)] . \tag{6.41}$$

As can be expected from the expressions in (6.25) and (6.41), this probability is larger than (6.25) for the ML solution.

6.5 ML and REML for an alternative formulation of the model

For some applications, the covariance in (6.3) is expressed as

$$\boldsymbol{\Sigma} = \sigma_p^2(\tau_1 \boldsymbol{V}_1 + \tau_2 \boldsymbol{V}_2 + \cdots + \tau_p \boldsymbol{V}_p) = \sigma_p^2 \boldsymbol{\Sigma}_0 , \tag{6.42}$$

where $\tau_p = 1$ and $\tau_i = (\sigma_i^2/\sigma_p^2)$ for $i = 1, 2, \ldots, p-1$. Replacing $\boldsymbol{\Sigma}$ in (6.6) by $\sigma_p^2 \boldsymbol{\Sigma}_0$, we find that the ML solutions for σ_p^2 and τ_i are given by

$$\hat{\sigma}_p^2 = \boldsymbol{Y}' \boldsymbol{R}_0 \boldsymbol{Y}/n \tag{6.43}$$

and

$$tr\boldsymbol{\Sigma}_0^{-1} \boldsymbol{V}_i = \boldsymbol{Y}' \boldsymbol{R}_0 \boldsymbol{V}_i \boldsymbol{R}_0 \boldsymbol{Y}/\hat{\sigma}_p^2, \tag{6.44}$$

where $\boldsymbol{R}_0 = \boldsymbol{\Sigma}_0^{-1}[\boldsymbol{I} - \boldsymbol{X}(\boldsymbol{X}'\boldsymbol{\Sigma}_0^{-1}\boldsymbol{X})^-\boldsymbol{X}'\boldsymbol{\Sigma}_0^{-1}]$. For the REML procedure, $\hat{\sigma}_p^2$ remains the same as (6.43) but the variance ratios τ_i are obtained from

$$tr\boldsymbol{R}_0 \boldsymbol{V}_i = \boldsymbol{Y}' \boldsymbol{R}_0 \boldsymbol{V}_i \boldsymbol{R}_0 \boldsymbol{Y}/\hat{\sigma}_p^2 . \tag{6.45}$$

6.6 Iterations and transformations

Iterative procedures have been suggested in the literature for the estimation of variance components and fixed effects from the ML and REML solutions. The steepest ascent, Newton–Raphson, scoring and EM algorithms, which are used in practice, are described in the following subsections.

In some situations, suitable transformations of the estimating equations can simplify the computations and they can be used along with the iterations. For instance, in the case of the one-way random effects model presented in Sections 6.3.2 and 6.4.1, the iterations may converge rapidly for σ_i^2 and $v_i = (\sigma_\alpha^2 + \sigma_i^2/n_i)$. The estimate of σ_α^2 in this case can be obtained from $\sum_i \hat{v}_i/k - (1/k)\sum_i \hat{\sigma}_i^2/n_i$.

To simplify the above type of iterations, Hemmerle and Hartley (1973) and Goodnight and Hemmerle (1973) suggested the W-transform. Jennerich and Sampson (1976), for instance, examined it for the ML estimation of the variance components of some of the random effects models. Callanan and Harville (1993) suggested several other transformations and found the Newton–Raphson and scoring algorithms to be satisfactory for speedy convergence of the estimates in the case of the one-way random effects model. Hartley, J.N.K. Rao and La Motte (1978) suggest an iterative procedure for estimating variance components.

6.6.1 Method of steepest ascent

Let $\hat{\sigma}_{i,t}^2$ and $\hat{\sigma}_{i,t+1}^2$ denote the estimates for σ_i^2 at the tth and $(t+1)$th stages of iteration. The *score* at the tth stage $d_{i,t}$ is defined as $\partial l/\partial \sigma_i^2$ evaluated at $\hat{\sigma}_{i,t}^2$. For this procedure,

$$\hat{\sigma}_{i,t+1}^2 = \hat{\sigma}_{i,t}^2 + c_t d_{i,t} \ . \tag{6.46}$$

where c_t is a suitably chosen constant. For the ML method, from (6.9)–(6.11), $d_{i,t} = (1/2)(g_i - tr\ \boldsymbol{WV}_i)_t$, and c_t is chosen such that $L(\hat{\sigma}_{i,t+1}^2; \mathbf{Y}) > L(\hat{\sigma}_{i,t}^2; \mathbf{Y})$. Similarly, for the REML procedure, $d_{i,t} = (1/2)(g_i - tr\boldsymbol{RV}_i)_t$ from (6.30) and (6.31), and c_t is chosen such that $L(\hat{\sigma}_{i,t+1}^2; \boldsymbol{A'Y}) > L(\hat{\sigma}_{i,t}^2; \boldsymbol{A'Y})$.

6.7 The Newton–Raphson algorithm

The estimate $\hat{\sigma}_{i,t+1}^2$ can be formally expressed as

$$\hat{\sigma}_{i,t+1}^2 = \hat{\sigma}_{i,t}^2 + (\hat{\sigma}_{i,t+1}^2 - \hat{\sigma}_{i,t}^2) = \hat{\sigma}_{i,t}^2 + \delta_{i,t} \ . \tag{6.47}$$

Now, as an approximation,

$$\left(\frac{\partial l}{\partial \sigma_i^2}\right)_{t+1} = \left(\frac{\partial l}{\partial \sigma_i^2}\right)_t + \delta_{it}\left(\frac{\partial^2 l}{\partial(\sigma_i^2)^2}\right)_t . \tag{6.48}$$

For the ML estimation, the derivative on the left hand side is set to zero. Hence,

$$\delta_{it} = -\left(\frac{\partial l}{\partial \sigma_i^2}\right)_t \div \left(\frac{\partial^2 l}{\partial(\sigma_i^2)^2}\right)_t . \tag{6.49}$$

The estimate at the $(t+1)$th stage is obtained by substituting this expression in (6.47). Similarly, for the REML method, δ_{it} is found from (6.49) with $\partial l_1/\partial \sigma_i^2$ and $\partial^2 l_1/\partial(\sigma_i^2)^2$.

For the case of $p(>1)$ variance components, let $\boldsymbol{d}_t$ denote the p-vector of $\partial l/\partial \sigma_i^2$ evaluated at $\hat{\sigma}_{it}^2$ and $\boldsymbol{D}_t$ the $p \times p$ matrix of $d_{ij} = \partial^2 l/\partial \sigma_i^2 \partial \sigma_j^2$ evaluated at $\hat{\sigma}_{i,t}^2$ and $\hat{\sigma}_{j,t}^2$. At the $(t+1)$th stage, the estimate of $\boldsymbol{\sigma}$ is obtained from

$$\widehat{\boldsymbol{\sigma}}_{t+1} = \widehat{\boldsymbol{\sigma}}_t + \boldsymbol{\delta}_t = \widehat{\boldsymbol{\sigma}}_t - \boldsymbol{D}_t^{-1}\boldsymbol{d}_t . \tag{6.50}$$

As an illustration, consider the one-way unbalanced design described in Section 6.3.2 and represent σ^2 and σ_α^2 by the subscripts (1,2). The vector of the first partial derivatives of $log_e L$ are $\boldsymbol{d}' = (d_1, d_2)$; notice that d_1 appears in (6.21) and d_2 in (6.20). The second partial derivatives are given by

$$d_{11} = \frac{n-k}{2\sigma^4} + \frac{1}{2}\sum_i \frac{1}{n_i^2 v_i^2} - \frac{\sum_i(n_i-1)s_i^2}{\sigma^6} - \sum_i \frac{(\bar{y}_i - \mu)^2}{n_i^2 v_i^3} ,$$

$$d_{22} = (1/2)\sum_i \frac{1}{v_i^2} - \sum_i \frac{(\bar{y}_i - \mu)^2}{v_i^3}, \text{ and}$$

$$d_{12} = (1/2)\sum_i \frac{1}{n_i v_i^2} - \sum_i \frac{(\bar{y}_i - \mu)^2}{n_i^2 v_i^3} . \tag{6.51}$$

Both $\boldsymbol{d}$ and $\boldsymbol{D}$ depend on the unknown parameters. The iterations can be started from (6.51), for instance, with $\hat{\mu} = \sum_i n_i\bar{y}_i/n, \hat{\sigma}^2 = \sum_i(n_i-1)s_i^2/(n-k)$ and $\hat{\sigma}_\alpha^2 = 0$.

6.8 Fisher's scoring method

In this approach, for the ML procedure, $\partial^2 l/\partial(\sigma_i^2)^2$ in (6.49) is replaced by its expectation evaluated at the tth stage, that is,

$-\mathcal{I}(\sigma_{i,t}^2)$. Note that the correction for the $(t+1)$th stage is obtained from the product of the *score* and the *information* at the tth stage. Thus, from (6.47) and (6.49),

$$\hat{\sigma}_{i,t+1}^2 = \hat{\sigma}_{i,t}^2 + d_{i,t}\mathcal{I}(\sigma_{i,t}^2) \; . \tag{6.52}$$

With the definitions for the information and score in Sections 6.3.1 and 6.6.1,

$$\hat{\sigma}_{i,t+1}^2 = \hat{\sigma}_{i,t}^2 + (g_i - tr\boldsymbol{W}\boldsymbol{V}_i)_t/(tr\boldsymbol{W}\boldsymbol{V}_i\boldsymbol{W}\boldsymbol{V}_j)_t \; . \tag{6.53}$$

For the case of p variance components,

$$\widehat{\boldsymbol{\sigma}}_{t+1} = \widehat{\boldsymbol{\sigma}}_t + \boldsymbol{H}_t^{-1}(\boldsymbol{g} - \boldsymbol{T})_t \; , \tag{6.54}$$

where $\boldsymbol{T}$ is the p-vector with the elements $tr\boldsymbol{W}\boldsymbol{V}_i$. Note that $\boldsymbol{H}_t$ is the matrix of $tr\;\boldsymbol{W}\boldsymbol{V}_i\boldsymbol{W}\boldsymbol{V}_j$ with $\hat{\sigma}_{i,t}^2$ substituted for σ_i^2.

For the REML procedure,

$$\hat{\sigma}_{i,t+1}^2 = \hat{\sigma}_{i,t}^2 + (g_i - tr\boldsymbol{R}\boldsymbol{V}_i)_t/tr(\boldsymbol{R}\boldsymbol{V}_i\boldsymbol{R}\boldsymbol{V}_j)_t \; . \tag{6.55}$$

For the case of more than one variance component,

$$\widehat{\boldsymbol{\sigma}}_{t+1} = \widehat{\boldsymbol{\sigma}}_t + \boldsymbol{F}_t^{-1}(\boldsymbol{g} - \boldsymbol{T}^*)_t \; , \tag{6.56}$$

where the p-vector $\boldsymbol{T}^*$ consists of the elements $tr\boldsymbol{R}\boldsymbol{V}_i\boldsymbol{R}\boldsymbol{V}_j$ with σ_i^2 replaced by $\hat{\sigma}_{i,t}^2$.

6.8.1 The EM algorithm

In this procedure, E and M refer to *expectation* and *maximization*. Laird (1982) showed that with this approach, the ML estimates are obtained from

$$n_i\hat{\sigma}_{i,t+1}^2 = \hat{\sigma}_{i,t}^2[n_i - \hat{\sigma}_{i,t}^2 tr(\boldsymbol{W}_t\boldsymbol{V}_i)] + \hat{\sigma}_{i,t}^4 g_{it} \; , \tag{6.57}$$

and noted that this expression was obtained by Hartley and J.N.K. Rao through suitable approximations. Estimates for the REML method are obtained from the same expression by replacing $tr(\boldsymbol{W}_t\boldsymbol{V}_i)$ with $tr(\boldsymbol{R}_t\boldsymbol{V}_i)$.

The second term of (6.57) provides the major contribution to the estimation of a variance component, and its significance will become more apparent in Chapter 8. Dempster, Rubin and Tsutakawa (1981) illustrated the application of the EM algorithm for estimating the variance and covariance components through the ML procedure.

6.9 Further investigations

Thompson (1962) considered both the row and column effects of the two-way classification model in (3.2) to be random. The REML solutions for the variance components of the balanced designs for this case coincided with the ANOVA estimators and were adjusted for nonnegativeness.

When the above balanced design is of the mixed type, Corbeil and Searle (1976) compared the ML and REML solutions of the variance components. With suitable adjustments for nonnegativeness, Lee and Kapadia (1984) compared the ML and REML estimators and found the ML estimator of the variance component to have a large negative bias.

The ML and REML estimators for the variance components of the three-stage nested classification were studied by Sahai (1976) for the balanced design and by P.S.R.S. Rao and Heckler (1997) for the unbalanced design. Miller (1979) and Brown and Burgess (1984) evaluated the performances of the ML and REML estimators through Monte Carlo studies.

For the balanced one-way random effects model, as presented in (2.17), the ANOVA estimator for σ_α^2/σ^2 is $(F-1)/m$, where $F = M_B/M_W$; an unbiased estimator was suggested in Exercise 2.7. For this ratio, the ML and REML estimators respectively are given by $[(k-1)F/k-1]^+/m$ and $(F-1)^+/m$, that is, adjustments to nonnegativeness for $\hat{\sigma}_\alpha^2$ are made as described earlier. Loh (1986) shows that $(cF-1)^+/m$, where $c = (k-1)(n-k-4)/(k+1)(n-k)$, has smaller MSE than the REML estimator, which in turn has smaller MSE than the ML estimator. The constant c is obtained by minimizing the MSE of cF for estimating $E(M_B)/E(M_W) = m\sigma_\alpha^2/\sigma^2 + 1$. For σ_α^2 and σ^2 of this balanced model, Klotz, Milton and Zacks (1969) suggested some modifications of the ML estimators.

Exercises

6.1 From the expression in (6.16) or directly from the one-way model in (2.1) and the normality assumption, show that the likelihood function can be expressed as (6.18).

6.2 Find the second partial derivatives of (6.19) with respect to μ, σ_i^2 and σ_α^2 and examine whether the solutions from (6.20) maximize the likelihood.

6.3 From the second derivatives in Exercise 6.2 find the information matrix and the Cramer–Rao lower bounds for the variances of the estimators of μ, σ_i^2 and σ_α^2.

6.4(a) From the logarithm of the likelihood in (6.19), for the case of $\sigma_i^2 \equiv \sigma^2$ and $n_i \equiv m$, derive the MLEs for μ, σ^2 and $\sigma_\alpha^2 + \sigma^2/m$, and show that the MLE for σ_α^2 is obtained from (6.23). (b) Find the information matrix for μ, σ^2 and $\sigma_\alpha^2 + \sigma^2/m$ and the Cramer–Rao lower bounds for the variances of the estimators of these parameters.

6.5 For the REML, obtain (6.38) from (6.31) and (6.32).

6.6 Consider the one-way random effects model with $\sigma_i^2 \equiv \sigma^2$ and $n_i \equiv m$. (a) From the logarithm of the likelihood in (6.36), find the REML estimators for σ^2 and $(\sigma_\alpha^2 + \sigma^2/m)$. (b) Find the information matrix for σ^2 and $(\sigma_\alpha^2 + \sigma^2/m)$ and the Cramer–Rao lower bounds for the variances of the estimators.

6.7 For the two-way classification model in (3.2), find the ML and REML estimators of the variance components and the residual variance when (a) both the rows and columns are random and (b) when the rows are fixed but the columns are random.

6.8 For the two-factor or three-stage design represented by (5.1), find the ML and REML estimators when both α_i and $\beta_{(i)j}$ are random.

CHAPTER 7

The MINQUE and MIVQUE

7.1 Introduction

Several procedures have been proposed for estimating the variance components of the model in (6.1) through quadratic forms in the observations. More than one unbiased estimator can be found for some of the variance components, and each of the estimators can have relatively smaller variance if certain conditions are satisfied. For instance, as seen in the case of the one-way unbalanced model, both the ANOVA and the USS estimators in (2.35) and (2.40) are unbiased for σ_α^2. A number of unbiased estimators can also be formed from linear combinations of these two estimators. However, the variances of each of such estimators depend on the number of groups, sample sizes in the groups and the magnitudes of $(\sigma_\alpha^2, \sigma^2)$. Henderson (1953) presents different methods for obtaining the ANOVA type of estimators.

C.R. Rao (1970, 1971a, 1971b, 1972) proposed the minimum norm quadratic unbiased estimation (MINQUE) which has desirable properties in addition to unbiasedness. It becomes the minimum variance quadratic unbiased estimator (MIVQUE) under suitable conditions. With the assumption of normality for the observations, La Motte (1973) describes procedures for quadratic estimation of variance components.

The MIVQUE can also utilize *a priori* values of the variance components, and iterative procedures of the type described in Chapter 6 are needed to find the MIVQUEs for unbalanced designs. The one-step iterated MIVQUE will be seen to be the same as the REML estimator.

C.R. Rao (1973, 1974, 1979) described further properties of the

MINQUE such as its closeness to the ML and REML estimators, and C.R. Rao and Kleffé (1989) comprehensively present this estimation procedure and its properties for the univariate and multivariate mixed models. P.S.R.S. Rao (1977) presents a review of the MINQUE.

The principle of the MINQUE has provided a renewed interest in the estimation of the variance and covariance components and an impetus for further research on the inference related to mixed models. Procedures based on the MINQUE for nonnegative estimation of variance components will be presented in Chapter 8.

7.2 Principle of the MINQUE

A linear combination of the variance components of the model in (6.1) is given by

$$l'\sigma = l_1\sigma_1^2 + l_2\sigma_2^2 + \cdots + l_p\sigma_p^2 , \tag{7.1}$$

where l is a $p \times 1$ row vector $(l_1, l_2, \cdots, l_p)$. The coefficients l_i can be suitably chosen to represent a single variance component or a specified linear combination of the variance components.

To estimate $l'\sigma$, consider a quadratic form of the observations $l'\widehat{\sigma} = Y'AY$, which is **invariant** to β if $AX = 0$. This condition implies that the quadratic form $(Y - X\beta_0)'A(Y - X\beta_0)$ for any $s \times 1$ vector β_0 remains the same as $Y'AY$. Note also that $AX = 0 \Rightarrow X'AX = 0$.

With this **translation invariance**,

$$Y'AY = \varepsilon'A\varepsilon = \xi'U'AU\xi \tag{7.2}$$

and

$$E(Y'AY) = Etr(AU\xi\xi'U') = \sum_i \sigma_i^2 tr AV_i . \tag{7.3}$$

Hence, $l'\widehat{\sigma}$ will be unbiased for $l'\sigma$ if $tr AV_i = l_i$ for $i = 1, 2, \ldots, p$.

If ξ_i is known, $\sum_1^{n_i} \xi_{ij}^2/n_i = \xi_i'\xi_i/n_i$ is an unbiased estimator for σ_i^2. To motivate the MINQUE principle for $l'\sigma$, C.R. Rao (1971) starts with the 'natural estimator'

$$(l_1/n_1)\xi_1'\xi_1 + (l_2/n_2)\xi_2'\xi_2 + \cdots + (l_p/n_p)\xi_p'\xi_p = \xi'\Delta\xi . \tag{7.4}$$

In this expression, Δ is a diagonal matrix with elements $(l_i/n_i)I_i$, where I_i is the identity matrix of dimension n_i.

From (7.2) and (7.4)

$$Y'AY - \xi'\Delta\xi = \xi'(U'AU - \Delta)\xi \ . \tag{7.5}$$

This difference may be reduced by minimizing the Euclidean norm

$$
\begin{aligned}
\|U'AU - \Delta\| &= tr(U'AU - \Delta)^2 \\
&= tr(AVAV + \Delta^2 - 2AU\Delta U') \\
&= trAVAV - tr\Delta^2 \ .
\end{aligned} \tag{7.6}
$$

Note that $tr\Delta^2 = \sum_i(l_i^2/n_i)$ and the last expression is obtained from the unbiasedness condition.

Thus, for the MINQUE of $l'\sigma$, the matrix A of $l'\hat{\sigma} = Y'AY$ is obtained by minimizing $trAVAV$ with the invariance and unbiasedness conditions

$$\text{(i)} \ \ AX = 0 \ \ \text{and} \ \ \text{(ii)} \ trAV_i = l_i \ . \tag{7.7}$$

7.2.1 The MINQUE solution

Note that

$$AX = 0 \Rightarrow BX_0 = 0 \ , \tag{7.8}$$

where $B = V^{1/2}AV^{1/2}$ and $X_0 = V^{-1/2}X$. The unbiasedness condition now becomes

$$trAV_i = trBV^{-1/2}V_iV^{-1/2} = l_i \ . \tag{7.9}$$

Thus, minimizing $trAVAV$ with the two conditions in (7.7) is the same as minimizing trB^2 with the conditions in (7.8) and (7.9). Now,

$$BX_0 = 0 \Rightarrow B = Q_0BQ_0 \ , \tag{7.10}$$

where $Q_0 = I - X_0(X_0'X_0)^-X_0'$. Hence,

$$trBV^{-1/2}V_iV^{-1/2} = trBQ_0V^{-1/2}V_iV^{-1/2}Q_0 = l_i \ . \tag{7.11}$$

The solution for minimizing trB^2 with the conditions in (7.10) and (7.11) is given by

$$B = \sum_i \lambda_i Q_0 V^{-1/2}V_iV^{-1/2}Q_0 \ . \tag{7.12}$$

Thus,

$$\boldsymbol{A} = \sum_i \lambda_i \boldsymbol{R}_0 \boldsymbol{V}_i \boldsymbol{R}_0 \ , \tag{7.13}$$

where $\boldsymbol{R}_0 = \boldsymbol{V}^{-1/2} \boldsymbol{Q}_0 \boldsymbol{V}^{-1/2} = \boldsymbol{V}^{-1}[\boldsymbol{I} - \boldsymbol{X}(\boldsymbol{X}'\boldsymbol{V}^{-1}\boldsymbol{X})^- \boldsymbol{X}'\boldsymbol{V}^{-1}]$. The coefficients λ_i are obtained from the unbiasedness condition in (7.7), that is, from

$$\sum_i \lambda_i tr \boldsymbol{R}_0 \boldsymbol{V}_i \boldsymbol{R}_0 \boldsymbol{V}_j = l_j \ , j = 1, 2, \ldots, p \tag{7.14}$$

This equation can be expressed as $\boldsymbol{F}_0 \boldsymbol{\lambda} = \boldsymbol{l}$, where $\boldsymbol{\lambda}$ and $\boldsymbol{l}$ are the vectors of λ_i and l_i respectively.

From (7.13), the MINQUE of $\boldsymbol{l}'\boldsymbol{\sigma}$ is given by

$$\boldsymbol{l}'\widehat{\boldsymbol{\sigma}} = \sum_i \lambda_i \boldsymbol{e}_0' \boldsymbol{V}_i \boldsymbol{e}_0 = \sum_i \lambda_i g_{0i} = \boldsymbol{\lambda}' \boldsymbol{g}_0 \ , \tag{7.15}$$

where $\boldsymbol{e}_0 = \boldsymbol{R}_0 \boldsymbol{Y}$, $g_{0i} = \boldsymbol{e}_0' \boldsymbol{V}_i \boldsymbol{e}$, and $\boldsymbol{g}_0$ is the vector of the elements g_{0i} for $i = 1, 2, \ldots, p$. Since $\boldsymbol{l}' = \boldsymbol{\lambda}' \boldsymbol{F}_0$, from (7.15)

$$\boldsymbol{F}_0 \widehat{\boldsymbol{\sigma}} = \boldsymbol{g}_0 \ . \tag{7.16}$$

Thus, the MINQUE's of the individual variance components and $\boldsymbol{l}'\boldsymbol{\sigma}$ are obtained from $\widehat{\boldsymbol{\sigma}} = \boldsymbol{F}_0^{-1} \boldsymbol{g}_0$ and $\boldsymbol{l}'\widehat{\boldsymbol{\sigma}} = \boldsymbol{l}' \boldsymbol{F}_0^{-1} \boldsymbol{g}_0$.

This procedure of estimating the variance components is equivalent to the method of equating $\boldsymbol{e}'\boldsymbol{V}_i \boldsymbol{e}$ to its expectation and solving for σ_i^2. As can be seen from (7.15) and (7.16), the MINQUE is a linear combination of the quadratic forms of the residuals $\boldsymbol{Q}_0 \boldsymbol{Y}$, obtained by regressing $\boldsymbol{V}^{-1/2}\boldsymbol{Y}$ on $\boldsymbol{V}^{-1/2}\boldsymbol{X}$.

7.2.2 Alternative derivations of the MINQUE

Alternative derivations have been suggested for the MINQUE of $\boldsymbol{l}'\boldsymbol{\sigma}$. To obtain $\boldsymbol{B}$ by minimizing $tr \boldsymbol{B}^2$ with the invariance and unbiasedness conditions in (7.10) and (7.11), following an instructive approach of C.R. Rao (1973), consider an alternative quadratic form $\boldsymbol{Y}'(\boldsymbol{B} + \boldsymbol{D})\boldsymbol{Y}$. With the additional conditions $\boldsymbol{DX}_0 = \boldsymbol{0}$ and $tr \boldsymbol{DV}^{-1/2} \boldsymbol{V}_i \boldsymbol{V}^{-1/2} = 0$, this quadratic form will be invariant to $\boldsymbol{\beta}$ and unbiased for $\boldsymbol{l}'\boldsymbol{\sigma}$. From the first condition,

$$\boldsymbol{DX}_0 = \boldsymbol{0} \Rightarrow \boldsymbol{D} = \boldsymbol{Q}_0 \boldsymbol{D} \boldsymbol{Q}_0 \ . \tag{7.17}$$

Hence,

$$tr \boldsymbol{DV}^{-1/2} \boldsymbol{V}_i \boldsymbol{V}^{-1/2} = tr \boldsymbol{D} \boldsymbol{Q}_0 \boldsymbol{V}^{-1/2} \boldsymbol{V}_i \boldsymbol{V}^{-1/2} \boldsymbol{Q}_0 = 0 \ . \tag{7.18}$$

This result implies that $tr\boldsymbol{B}\boldsymbol{D} = 0$ if $\boldsymbol{B}$ takes the form of (7.12). As a result,

$$tr(\boldsymbol{B} + \boldsymbol{D})^2 = tr\boldsymbol{B}^2 + tr\boldsymbol{D}^2 \geq tr\boldsymbol{B}^2 \; . \tag{7.19}$$

Thus, the optimum $\boldsymbol{B}$ and hence $\boldsymbol{A}$ are given by (7.12) and (7.13) respectively.

For another approach suggested by C.R. Rao (1984), it is noted that

$$\boldsymbol{B}\boldsymbol{X}_0 = 0 \Rightarrow \boldsymbol{B} = \boldsymbol{Z}\boldsymbol{C}\boldsymbol{Z}' \; , \tag{7.20}$$

where $\boldsymbol{Z}$ is an $n\times(n-r)$ matrix of rank $(n-r)$ such that $\boldsymbol{Z}\boldsymbol{Z}' = \boldsymbol{Q}_0$ and $\boldsymbol{Z}'\boldsymbol{Z} = \boldsymbol{I}$, and the $(n-r)\times(n-r)$ matrix $\boldsymbol{C}$ is of full rank. Now, minimizing $tr\boldsymbol{B}^2$ with the invariance and unbiasedness conditions in (7.10) and (7.11) is the same as minimizing $tr\boldsymbol{C}^2$ with

$$tr\boldsymbol{C}(\boldsymbol{Z}'\boldsymbol{V}^{-1/2}\boldsymbol{V}_i\boldsymbol{V}^{-1/2}\boldsymbol{Z}) = l_i \; . \tag{7.21}$$

The solution for this optimization is given by

$$\boldsymbol{C} = \sum_i \lambda_i(\boldsymbol{Z}'\boldsymbol{V}^{-1/2}\boldsymbol{V}_i\boldsymbol{V}^{-1/2}\boldsymbol{Z}) \; . \tag{7.22}$$

Consequently $\boldsymbol{B}$ and $\boldsymbol{A}$ are the same as (7.12) and (7.13).

Mitra (1971) derived (7.16) by considering quadratic forms in $\boldsymbol{Y}$ whose expectations are linear functions of the variance components. Brown (1977) derived the MINQUE from the weighted least squares residuals.

The computations required for the MINQUE may be simplified by considering a suitable nonsingular transformation $\boldsymbol{Z} = \boldsymbol{H}'\boldsymbol{Y}$ and the corresponding quadratic form in $\boldsymbol{Z}$. C.R. Rao (1970) demonstrates the advantage of selecting an orthogonal matrix for $\boldsymbol{H}$. As suggested by C.R. Rao (1971a), the computations can also be simplified if $\boldsymbol{H}$ is an $n \times (n - r)$ matrix of rank $(n - r)$ such that $\boldsymbol{H}'\boldsymbol{X} = \boldsymbol{0}$.

7.3 MINQUE with *a priori* values

With γ_i^2 denoting the *a priori* values of σ_i^2 for $i = 1, 2, \ldots, p$, for the model in (6.1),

$$\varepsilon = \boldsymbol{U}_{1*}\boldsymbol{\eta}_1 + \boldsymbol{U}_{2*}\boldsymbol{\eta}_2 + \cdots + \boldsymbol{U}_{p*}\boldsymbol{\eta}_p = \boldsymbol{U}_*\boldsymbol{\eta} \; , \tag{7.23}$$

where $U_{i*} = U_i\gamma_i$, $U_* = (U_{1*}, U_{2*}, \ldots, U_{p*})$, $\eta_i = (1/\gamma_i)\xi_i$ and $\eta' = (\eta'_1, \eta'_2, \ldots, \eta'_p)$. Note that $U_* = U\Lambda^{1/2}$ and $\eta = \Lambda^{-1/2}\xi$, where $\Lambda^{1/2}$ is a diagonal matrix with elements $\gamma_i\, I_i$.

From (7.23), with the invariance condition $AX = 0$,

$$Y'AY = \xi'U'AU\xi = \eta'(\Lambda^{1/2}U'AU\Lambda^{1/2})\eta , \tag{7.24}$$

and the natural unbiased estimator $\sum_i (l_i/n_i)\, \xi'_i\xi_i$ in (7.4) can be expressed as

$$\xi'\Delta\xi = \eta'\Lambda^{1/2}\Delta\Lambda^{1/2}\eta . \tag{7.25}$$

To minimize the difference between (7.24) and (7.25), consider

$$tr[\Lambda^{1/2}(U'AU - \Delta)\Lambda^{1/2}]^2$$
$$= tr A(U\Lambda U')A(U\Lambda U') + tr\Delta\Lambda\Delta\Lambda - 2tr AU(\Lambda\Delta\Lambda)U'$$
$$= tr AV_*AV_* - \Sigma(l_i/n_i)\gamma_i^2 , \tag{7.26}$$

where $V_* = U_*U'_*$. The last expression is obtained through the unbiasedness condition.

The MINQUE for $l'\sigma$ is obtained by minimizing $tr AV_*AV_*$ with the conditions in (7.7). Through any of the optimization procedures in Sections (7.2.1) and (7.2.2), the solution is given by

$$A = \sum_i \lambda_i RV_iR , \tag{7.27}$$

where $Q_* = I - X(X'V_*^{-1}X)^- X'V_*^{-1}$ and $R = V_*^{-1}Q_*$. The coefficients λ_i are obtained from

$$\sum_i \lambda_i tr RV_iRV_j = l_j , j = 1, 2, \ldots p . \tag{7.28}$$

As before,

$$l'\hat{\sigma} = \sum_i \lambda_i e'V_ie = \sum_i \lambda_i g_i , \tag{7.29}$$

where $e = RY$ and $g_i = e'V_ie$. The MINQUEs of σ_i^2 are obtained from

$$F\hat{\sigma} = g . \tag{7.30}$$

The elements of F are given by $f_{ij} = tr RV_iRV_j$ and the vector g consists of the elements g_i.

As noted in Section 7.2.1, this procedure of estimation is the same as equating $e'V_ie$ to its expectation and solving for σ_i^2.

As noted in Section 6.4, the REML estimators are obtained by the same procedure as above except that V_* in R, e, F and g is replaced by Σ and the constraints for nonnegativeness are imposed.

Simplifications for finding the MIVQUEs from (7.16) or (7.30) for special cases were suggested by Lou and Senturia (1977), P.S.R.S. Rao, Kaplan and Cochran (1981), Kaplan (1983), Giesbrecht (1983), Kleffé and Siefert (1986), and others.

7.4 Further properties of the MINQUE

In addition to the invariance to the fixed parameters and the unbiasedness, the MINQUE has other desirable properties.

(i) When all the n_i elements of ξ_i have the same variance σ_i^2 and the same fourth moment μ_{4i}, with the invariance condition $AX = 0$,

$$V(Y'AY) = 2tr A\Sigma A\Sigma + \sum_i (\mu_{4i} - 3\sigma_i^4) tr AV_i AV_i . \qquad (7.31)$$

The second term in this expression vanishes, for instance, if ξ_i follows a normal distribution. In this case, the MINQUE derived in Section 7.3 is the same as the estimator obtained by minimizing (7.31) with $V_* = \sum_i \gamma_i^2 V_i$ in the place of Σ. Thus, the MINQUE becomes the same as the MIVQUE with *a priori* values.

(ii) The MINQUE in (7.29) is a linear combination of the quadratic forms of the residuals Q_*Y, which are obtained by regressing $V_*^{-1/2}Y$ on $V_*^{-1/2}X$.

(iii) If γ_i^2 are close to σ_i^2 for $i = 1, 2, \ldots, p$, the REML estimates and MINQUEs for σ_i^2 obtained from (6.32) and (7.30), with the constraints for nonnegativeness in both cases, can be expected to be the same.

(iv) When γ_i^2 are not available, the MINQUEs for σ_i^2 can be obtained from (7.30) through the iterative procedures in Section 6.8. As can be seen, a one-step iterated MIVQUE with the constraints for nonnegativeness is the same as the REML obtained from (6.32).

7.5 Illustrations and investigations of the MIVQUE

7.5.1 Common mean

The model

$$y_{ij} = \mu + \varepsilon_{ij} \ , \tag{7.32}$$

$i = 1, 2, \ldots, k$ and $j = 1, 2, \ldots, n_i$, with $E(\varepsilon_{ij}) = 0$ and $V(\varepsilon_{ij}) = \sigma_i^2$, represents observations from samples of sizes n_i from k populations with a common mean μ but different variances σ_i^2. The residuals ε_{ij} within a group and among the groups are assumed to be uncorrelated. For the case of $n_i \equiv m$, Cochran (1937) and Cochran and Carroll (1953) examined the estimation of μ and the standard error of its estimator through the maximum likelihood, weighted least squares (WLS) and related procedures.

With the notation in Section 7.3, for the MIVQUEs of σ_i^2 of this model, we find that

$$\gamma_i^4 e' V_i e = (n_i - 1)s_i^2 + n_i(\overline{y}_i - \overline{y}_W)^2 \ , \tag{7.33}$$

where $\overline{y}_i$ and s_i^2 are the sample mean and variance of the ith group, $W_i = n_i/\gamma_i^2, W = \sum_i W_i$ and $\overline{y}_W = \sum_i W_i \overline{y}_i/W$. From (7.33),

$$\gamma_i^4 E(e' V_i e) = (n_i - 2W_i/W)\sigma_i^2 + (n_i/W^2) \sum_i W_i^2 \sigma_i^2/n_i \ . \tag{7.34}$$

The MIVQUEs are obtained by equating the right hand sides of (7.33) and (7.34) and solving for σ_i^2. Alternatively, the estimates can be obtained from (7.30) with

$$f_{ii} = (n_i - 1) + (1 - W_i/W)^2 \ ,$$
$$f_{ij} = n_i W_j^2/n_j W^2$$

and

$$g_i = (n_i - 1)s_i^2 + n_i(\overline{y}_i - \overline{y}_W)^2 \ . \tag{7.35}$$

As can be seen from (7.34) or (7.35), the MIVQUEs of σ_i^2 depend only on the relative values of the *a priori* values. In the absence of any information regarding the variances, all the γ_i^2 or their relative values may be considered to be equal to a nonzero constant, unity in particular. This procedure will be the same as estimating σ_i^2 from (7.16). With this approach, from (7.33) and (7.34),

$$\hat{\sigma}_i^2 = \frac{1}{n-2} \left[\frac{n}{n_i} \sum_j (y_{ij} - \overline{y})^2 - \frac{\sum_i \sum_j (y_{ij} - \overline{y})^2}{n-1} \right] \ , \tag{7.36}$$

where $n = \sum_i n_i$ and $\overline{y} = \sum_i \sum_j y_{ij}/n$.

J.N.K. Rao (1973) examined the WLS estimator $\overline{y}_W = \sum_i W_i \overline{y}_i / W$ for μ of (7.32) with the weights $n_i/\hat{\sigma}_i^2$ obtained from (7.37). If $\hat{\sigma}_i^2$ was negative, it was replaced by s_i^2, a small positive quantity or the average of the squared residuals (ASR) given by $\sum_j e_{ij}^2/n_i = \sum_j (y_{ij} - \overline{y}_W)^2/n_i$. Chaubey and P.S.R.S. Rao (1976) further investigated the estimation of μ with additional modifications for nonnegativeness of the estimates. Both these studies found that the variance of the MIVQUE and the MSE of the ASR, in general, are smaller than the variance of s_i^2. The corresponding WLS estimators of μ were also found to have smaller MSEs.

With the expressions in Section 6.4, the REMLs for σ_i^2 are obtained from

$$n_i(w\sigma_i^2 - 1)/w\sigma_i^2 = (n_i - 1)s_i^2 + n_i(\overline{y}_i - \overline{y}_w)^2 \ , \tag{7.37}$$

where $w_i = n_i/\sigma_i^2$, $w = \sum_i w_i$ and $\overline{y}_w = \sum_i w_i \overline{y}_i/w$. This equation can also be obtained from (7.33) and (7.34) by replacing γ_i^2 with σ_i^2. If $\sigma_i^2 \equiv \sigma^2$, for both the MIVQUE and REML procedures, $\hat{\sigma}^2 = \sum_i \sum_j (y_{ij} - \overline{y})^2/(n-1)$.

7.5.2 Regression model

The regression of y_i on x_i with unequal variances and with replications is represented by

$$y_{ij} = \alpha + \beta x_i + \varepsilon_{ij} \ , \tag{7.38}$$

$i = 1, 2, \ldots, k$ and $j = 1, 2, \ldots, n_i$. The assumptions regarding variances and covariances of ε_{ij} are the same as described in Section 7.5.1. Let

$$b = \frac{\sum_i W_i(x_i - \overline{x}_W)(\overline{y}_i - \overline{y}_W)}{\sum_i W_i(x_i - \overline{x}_W)^2} \tag{7.39}$$

where $\overline{x}_W = \sum_i W_i x_i / W$. For the MIVQUE,

$$\gamma_i^4 e' V_i e = (n_i - 1)s_i^2 + n_i[(\overline{y}_i - \overline{y}_W) - b(x_i - \overline{x}_W]^2 \ . \tag{7.40}$$

The estimates of σ_i^2 are obtained by equating the right hand side to its expectation.

Jacqez, Mather and Crawford (1968) examined the least squares and maximum likelihood methods for estimating σ_i^2 and (α, β).

The investigations of J.N.K. Rao and Subrahmanium (1971) and Chaubey and P.S.R.S. Rao (1976) brought out the advantages of the MINQUE and ASR for this purpose.

7.5.3 One-way random effects model

For this model described in Section 2.8, the MIVQUEs of $(\sigma_\alpha^2, \sigma_i^2)$ are obtained from

$$g_\alpha = e'V_\alpha e = \sum_i W_i^2(\bar{y}_i - \bar{y}_W)^2 \tag{7.41}$$

and

$$g_i = e'V_i e = (n_i - 1)s_i^2/\gamma_i^4 + W_i^2(\bar{y}_i - \bar{y}_W)^2/n_i , \tag{7.42}$$

$i = 1, 2, \ldots, k$, where $W_i = n_i/(n_i\gamma_\alpha^2 + \gamma_i^2)$, $W = \sum_i W_i$ and $\bar{y}_W = \sum_i W_i\bar{y}_i/W$. Note that W_i is the reciprocal of $v_i = V(\bar{y}_i) = (\sigma_\alpha^2 + \sigma_i^2/n_i)$ with $(\gamma_\alpha^2, \gamma_i^2)$ substituted for $(\sigma_\alpha^2, \sigma_i^2)$. The expectations of (g_α, g_i) can be found from noting that $E(s_i^2) = \sigma_i^2$ and

$$E(\bar{y}_i - \bar{y}_W)^2 = v_i + \sum_i W_i^2 v_i/W^2 - 2W_i v_i/W . \tag{7.43}$$

To obtain the MIVQUEs from (7.30), the elements of the $(k + 1) \times (k + 1)$ matrix F for $(i \neq j) = 1, 2, \ldots, k$ are given by

$$f_{\alpha\alpha} = \sum_i W_i^2[1 + \sum_i (W_i/W)^2 - 2(W_i/W)] ,$$

$$f_{\alpha i} = (W_i^2/n_i)[1 + \sum_i (W_i/W)^2 - 2(W_i/W)] ,$$

$$f_{ii} = (n_i - 1)/\gamma_i^4 + (W_i/n_i)^2(W_i - W)^2/W^2$$

and

$$f_{ij} = W_i^2 W_j^2/n_i n_j W^2 . \tag{7.44}$$

If $\sigma_i^2 \equiv \sigma^2$, γ_i^2 is replaced by the common *a priori* value γ^2, and g_α remains the same as (7.42) with $W_i = n_i/(n_i\gamma_\alpha^2 + \gamma^2)$. Denoting σ^2 by the subscript zero, from (7.43),

$$g_0 = \sum_i g_i = \sum_i (n_i - 1)s_i^2/\gamma^4 + \sum_i W_i^2(\bar{y}_i - \bar{y}_W)^2/n_i . \tag{7.45}$$

The MIVQUEs of $(\sigma_\alpha^2, \sigma^2)$ are found from (7.42) and (7.46). The elements of F for this case are given by

$$f_{\alpha\alpha} = \sum_i W_i^2 - 2\sum_i W_i^2/W + (\sum_i W_i^2)^2/W^2 \;,$$

$$f_{\alpha 0} = \sum_i W_i^2/n_i - 2\sum_i W_i^3/n_i W + (\sum_i W_i^2)(\sum_i W_i^2/n_i W^2)$$

and

$$f_{00} = \frac{(n-k)}{\gamma^4} + \sum_i \frac{W_i^2}{n_i^2} - 2\sum_i \frac{W_i^3}{n_i^2 W} + \left(\sum_i \frac{W_i^2}{n_i W}\right)^2 \;. \quad (7.46)$$

If $\sigma_i^2 \equiv \sigma^2$ and $n_i \equiv m$, the MIVQUEs of $\sigma_\alpha^2, \sigma^2$ become the same as the ANOVA estimators in Section 2.3.1. If γ_α^2 is large relative to γ_i^2 or n_i are large, the MIVQUE of σ_α^2 becomes the same as the USS estimator in (2.40).

P.S.R.S. Rao, Kaplan and Cochran (1981) derived (7.41) and (7.42) and compared the MIVQUEs with the ANOVA and other procedures. It was found that for estimating $(\sigma_\alpha^2, \sigma_i^2)$ the variance of the MIVQUE remains smaller than the variance of the ANOVA estimator, even when γ_i^2 deviates slightly from σ_i^2. A similar result was found for the WLS estimator of μ with estimated weights.

Further investigations of the MIVQUEs of the one-way model were made by Swallow and Searle (1978), Hess (1979), Swallow (1981), P.S.R.S. Rao (1982), Swallow and Monahan (1984), P.S.R.S. Rao and Sylvestre (1984), and others. Westfall (1987) compares the ANOVA estimators and MIVQUEs of $(\sigma_\alpha^2, \sigma_i^2)$ for large n_i and nonnormal distributions.

7.5.4 Regression with a random intercept

Consider the model

$$y_{ij} = \delta_i + \beta x_i + \varepsilon_{ij} = \delta + \alpha_i + \beta x_i + \varepsilon_{ij} \;, \quad (7.47)$$

$i = 1, 2, \ldots, k$ and $j = 1, 2, \ldots, n_i$, where the intercept δ_i is a random variable with mean δ and variance σ_α^2. The assumptions regarding α_i and ε_{ij} are the same as in Section 2.8 for the one-way random effects model.

For the MIVQUE procedure,

$$g_\alpha = \sum_i W_i^2[(\bar{y}_i - \bar{y}_W) - b(x_i - \bar{x}_W)]^2 \quad (7.48)$$

and

$$g_i = (n_i - 1)s_i^2/\gamma_i^4 + (W_i^2/n_i)[(\overline{y}_i - \overline{y}_W) - b(x_i - \overline{x}_W)]^2 \ . (7.49)$$

The weights W_i are given by $n_i/(n_i\gamma_\alpha^2 + \gamma_i^2)$ as defined in Section 7.5.3 and the regression coefficient b takes the form of (7.39) with these weights.

The MIVQUEs of $(\sigma_\alpha^2, \sigma_i^2)$ are obtained from equating (7.48) and (7.49) to their expected values. If $\sigma_i^2 \equiv \sigma^2$ and $n_i \equiv m$, the MIVQUE of σ^2 is given by $\hat{\sigma}^2 = \sum_i s_i^2/k$ and of σ_α^2 by

$$\hat{\sigma}_\alpha^2 = \frac{\sum_i[(\overline{y}_i - \overline{y}) - b(x_i - \overline{x})]^2}{k - 2} - \frac{\hat{\sigma}^2}{m} \ . \tag{7.50}$$

In this expression $\overline{x} = \sum_i x_i/k, \overline{y} = \sum_i \overline{y}_i/k$, and $b = \sum_i(x_i - \overline{x})(\overline{y}_i - \overline{y})/\sum_i(x_i - \overline{x})^2$ is the ordinary least squares estimator.

P.S.R.S. Rao and Kuranchie (1988) derived (7.48) and (7.49), and compared the MIVQUEs of $(\sigma_\alpha^2, \sigma_i^2)$ and the corresponding WLS estimators of (δ, β) with other types of estimators. Kackar and Harville (1984) developed approximations to the S.E.s of the estimators of fixed and random effects of the above type of models.

7.5.5 One-way model with a covariate

This model with n_i observations from the kth group is represented by

$$y_{ij} = \delta_i + \beta x_{ij} + \varepsilon_{ij} = \delta + \alpha_i + \beta x_{ij} + \varepsilon_{ij} \ , \tag{7.51}$$

$i = 1, 2, \ldots, k$ and $j = 1, 2, \ldots, n_i$. Th error ε_{ij} is assumed have mean zero and variance σ_i^2. The random effect α_i is assumed to be independent of ε_{ij} with mean zero and variance σ_α^2.

Following the notation in Section 7.5.4, let

$$B_{xx} = \sum_i W_i(\overline{x}_i - \overline{x}_W)^2, B_{xy} = \sum_i(\overline{x}_i - \overline{x}_W)(\overline{y}_i - \overline{y}_W) \ ,$$

$$E_{xx} = \sum_i \sum_j(x_{ij} - \overline{x}_i)^2/\gamma_i^2$$

and

$$E_{xy} = \sum_i \sum_j(x_{ij} - \overline{x}_i)(y_{ij} - \overline{y}_i)/\gamma_i^2 \ . \tag{7.52}$$

Further, let

$$b = \frac{B_{xy} + E_{xy}}{B_{xx} + E_{xx}} \ . \tag{7.53}$$

This expression is the same as the WLS estimator of β with $(\sigma_\alpha^2, \sigma_i^2)$ replaced by $(\gamma_\alpha^2, \gamma_i^2)$. With these two expressions,

$$g_\alpha = \sum_i W_i^2 [(\bar{y}_i - \bar{y}_W) - b(\bar{x}_i - \bar{x}_W)]^2 \tag{7.54}$$

and

$$g_i = \sum_j [(y_{ij} - \bar{y}_i) - b(x_{ij} - \bar{x}_i)]^2 / \gamma_i^4$$

$$+ \sum_i [(\bar{y}_i - \bar{y}_W) - b(\bar{x}_i - \bar{x}_W)]^2 / n_i \ . \tag{7.55}$$

The MIVQUEs of $(\sigma_\alpha^2, \sigma_i^2)$ are obtained by equating (7.55) and (7.56) to their expectations. For the case of $\sigma_i^2 = \sigma^2$ and $n_i = m$, P.S.R.S. Rao and Miyawaki (1989) compared the MIVQUEs of $(\sigma_\alpha^2, \sigma^2)$ and the resulting extimator of β with the ANOVA, ML and other types of estimators. With the additional assumption that $\delta = 0$, Maddala and Mount (1973) compared the estimators for β obtained through the MIVQUE and other procedures.

7.5.6 Instrument precision and product variability

Consider observations on a sample of n units obtained from each of p instruments such as X-ray machines. These observations can be represented by the model

$$y_{ik} = \mu_i + \alpha_k + \varepsilon_{ik} \ , \tag{7.56}$$

for $i = 1, 2, \ldots, p$ and $k = 1, 2, \ldots n$. For the ith instrument, the mean is denoted by μ_i and the residual ε_{ik} is assumed to have mean zero and variance σ_i^2. The random effect α_k is assumed to be independent of ε_{ik} with mean zero and variance σ_α^2.

The sample means and variances of the instruments are $\bar{y}_i = \sum_k y_{ik}/n$ and $s_i^2 = \sum_k (y_{ik} - \bar{y}_i)^2/(n-1)$ for $i = 1, 2, \ldots p$, and the sample covariances are $s_{ij} = \sum_k (y_{ik} - \bar{y}_i)(y_{jk} - \bar{y}_j)/(n-1)$ for $(i = j) = 1, 2, \ldots, p$.

When the *a priori* values of $(\sigma_\alpha^2, \sigma_i^2)$ are all the same, from (7.30), the MIVQUEs of σ_α^2 and σ_i^2 are given by

$$\hat{\sigma}_\alpha^2 = \sum_i \sum_j s_{ij} / p(p-1) \tag{7.57}$$

and

$$\hat{\sigma}_i^2 = s_i^2 + \sum_i \sum_j s_{ij}/p(p-1) - 2\sum_j s_{ij}/(p-1) . \qquad (7.58)$$

As can be seen from (7.58), $\hat{\sigma}_1^2 = s_{11} - s_{12}$ and $\hat{\sigma}_2^2 = s_{22} - s_{12}$ for $p = 2$. For $p = 3$, $\hat{\sigma}_1^2 = s_1^2 - (2/3)(s_{12} + s_{13}) + (1/3)s_{23}$, and the estimators for σ_2^2 and σ_3^2 take similar forms. Grubbs (1948) suggested the same estimators for $p = 2$, but for $p > 2$ his estimators are obtained by replacing the denominators of the second and third terms on the right hand side of (7.58) by $(p-1)(p-2)$ and $(p-2)$; his estimator for σ_α^2 is the same as (7.57).

7.6 Iterated MIVQUE

To find the MIVQUEs from (7.30), the γ_i^2 or their relative values can be initially specified, assumed equal or estimated through a simple procedure like the ANOVA. The resulting estimates of σ_i^2 can be used as the *a priori* values for the second stage MIVQUEs. This iterative procedure can be continued until the differences in the estimates at the successive stages become negligible. Note that if γ_i^2 or their relative values are assumed to be equal, the first stage MIVQUEs obtained from (7.16) and (7.30) will be the same.

For the illustration of the one-way model presented in Sections 2.6 and 2.7, the ANOVA estimates were $\hat{\sigma}^2 = s^2 = 1.0435$ and $\hat{\sigma}_\alpha^2 = 0.9081$; s^2 was truncated to 1.04. The unweighted sum of squares (USS) estimate as found in Section 2.7.3 was equal to 1.1084.

The MIVQUEs of $(\sigma_\alpha^2, \sigma^2)$ and the corresponding WLS estimate $\bar{y}_W$ with estimated weights are presented in Table 7.1 for three iterations. For $(\gamma_\alpha^2, \gamma^2)$, the above two types of estimates and (1,1), (1,2) and (2,1) were considered. Note that the estimates depend only on $(\gamma_\alpha^2/\gamma^2)$ but not on the individual values of γ_α^2 and γ^2. The rapid change of the estimates at the first stage of iteration and the quick convergence to stable values from the second stage onwards should be recognized.

7.7 Covariance components

For the model in (6.1), let $n_i \equiv q$ and consider $\boldsymbol{\xi}_i$ to have zero expectations with $E(\boldsymbol{\xi}_i\boldsymbol{\xi}_i') = \boldsymbol{\Phi}$ and $E(\boldsymbol{\xi}_i\boldsymbol{\xi}_j') = \mathbf{0}$ for $(i \neq j) =$

Table 7.1. *Iterated MIVQUEs of $(\sigma_\alpha^2, \sigma^2)$ and the WLS estimate of μ*

γ_α^2	γ^2	$\hat{\sigma}_\alpha^2$	$\hat{\sigma}^2$	$\overline{y}_W$
$\hat{\sigma}_{\alpha A}^2$	s^2	0.6755	1.0457	2.3543
		0.6756	1.0470	2.3598
		0.6535	1.0470	2.3604
		0.6533	1.0472	2.3605
$\hat{\sigma}_{\alpha U}^2$	s^2	0.6869	1.0451	2.3512
		0.6568	1.0470	2.3595
		0.6536	1.0472	2.3605
		0.6533	1.0472	2.3605
1	1	0.6836	1.0453	2.3521
		0.6564	1.0470	2.3596
		0.6533	1.0472	2.3604
		0.6530	1.0472	2.3605
1	2	0.6362	1.0486	2.3653
		0.6509	1.0474	2.3611
		0.6527	1.0472	2.3606
		0.6529	1.0472	2.3606
2	1	0.7148	1.0457	2.3437
		0.6595	1.0468	2.3587
		0.6537	1.0472	2.3603
		0.6530	1.0471	2.3605

The ANOVA and USS estimators are $\hat{\sigma}_{\alpha A}^2 = 0.9081$ and $\hat{\sigma}_{\alpha U}^2 = 1.1084$, and $s^2 = 1.0435$.

$1, 2, \ldots, p$. With the invariance condition $\boldsymbol{AX} = \boldsymbol{0}$, the dispersion of $\boldsymbol{Y}$ becomes

$$\boldsymbol{\Sigma} = \sum_{i=1}^{p} \boldsymbol{U}_i \boldsymbol{\Phi} \boldsymbol{U}_i' \; . \tag{7.59}$$

Let $\boldsymbol{\Sigma}_*$ denote the *a priori* value of this dispersion. Further, let $\boldsymbol{Q}_* = \boldsymbol{I} - \boldsymbol{X}(\boldsymbol{X}'\boldsymbol{\Sigma}_*^{-1}\boldsymbol{X})^{-}\boldsymbol{X}'\boldsymbol{\Sigma}_*^{-1}$, $\boldsymbol{R} = \boldsymbol{\Sigma}_*^{-1}\boldsymbol{Q}_*$ and $\boldsymbol{e} = \boldsymbol{RY}$.

As noted at the end of Section 7.3, the MINQUEs of the variance components are obtained from $\boldsymbol{e}'\boldsymbol{V}_i\boldsymbol{e} = (\boldsymbol{e}'\boldsymbol{U}_i)(\boldsymbol{e}'\boldsymbol{U}_i)'$ where the residual $\boldsymbol{e}$ is obtained with $\boldsymbol{V}_*$. Similarly, with the above definition for $\boldsymbol{e}$,

$$E(\boldsymbol{U}_i'\boldsymbol{e})(\boldsymbol{U}_i'\boldsymbol{e})' = \boldsymbol{U}_i'\boldsymbol{R}(\textstyle\sum_i \boldsymbol{U}_i \boldsymbol{\Phi} \boldsymbol{U}_i')\boldsymbol{R}\boldsymbol{U}_i \tag{7.60}$$

for $i = 1, 2, \ldots, p$. Removing the expectation sign, and summing over $i = 1, 2, \ldots, p$, the estimator for $\boldsymbol{\Sigma}$ is obtained from

$$\sum_{1}^{p} \boldsymbol{U}_i'\boldsymbol{e}\boldsymbol{e}'\boldsymbol{U}_i = \sum_{j}\sum_{i} \boldsymbol{U}_j'\boldsymbol{R}\boldsymbol{U}_i \boldsymbol{\Sigma} \boldsymbol{U}_i'\boldsymbol{R}\boldsymbol{U}_j \; . \tag{7.61}$$

C.R. Rao (1972) obtains this estimator from the MINQUE principle and extends the procedure to models representing both the variance and covariance components. Chew (1971) obtains the estimator for $\boldsymbol{\Phi}$ from (7.62) without the *a priori* values, that is considering $\boldsymbol{\Sigma}_*$ to be the identity matrix.

7.8 Further applications of the MIVQUE

1. Designs of experiments

The MIVQUE was examined by Shah and Puri (1976) for the imcomplete block design and by Bremer (1990) for a factorial design with two factors. For the three-stage nested design, P.S.R.S. Rao and Heckler (1997) compared the MIVQUEs with the ML and ANOVA estimators and a procedure of the type described in Exercise 7.5.

2. Sample survey errors

Errors made by interviewers and other types of errors in a sample survey can be included in a model for analyzing its data. For such a model, Kleffé, Prasad and J.N.K. Rao (1991) compared the MIVQUEs with other estimators.

3. Correlations

For models of the type in (7.39), Chaubey (1980) considers $V(\varepsilon_{ijk})$ $= \sigma^2$ and $Cov(\varepsilon_{ij}, \varepsilon_{ij'}) = \rho\sigma^2$ for $(j \neq j')$, and obtains the estimator of the intraclass correlation ρ from the MIVQUEs of σ^2 and $\rho\sigma^2$. Note that ρ can be estimated from the model in (7.47) by expressing σ_α^2 as $\rho\sigma^2$; see also Section 2.4. This type of structure for the variances and covariances was described by Wirkowski (1975) for a regression model and by Wolfinger (1993) for mixed models. Eliasziw and Donner (1990) compared the different estimators suggested in the literature for the correlations of traits between parents and offspring and between offspring. Kleffé (1993) derives estimators for these correlations through the MINQUE procedure.

4. Variograms

Marshal and Mardia (1985) and Stein (1987) derive the MIVQUEs of variograms, which are correlations of random processes.

Exercises

7.1 Derive the expressions in (7.33)–(7.35).

7.2 Derive the MIVQUE for σ_i^2 in (7.36).

7.3 Derive the expression in (7.40) for the regression model.

7.4 Derive the expressions in (7.41) and (7.42) for the one-way random effects model.

7.5 Consider the one-way unbalanced random effects model with unequal residual variances. With arbitrary weights W_i, $W = \sum_i W_i$, show that (a) an unbiased estimator for σ_α^2 is given by $\sum_i [W_i(\bar{y}_i - \bar{y}_W)^2 - \sum_i W_i(W - W_i)s_i^2/n_i]/(W - \sum_i W_i^2/W)$, and (b) this estimator becomes the same as the ANOVA estimator for the balanced model with equal residual variances.

7.6 Show that for the balanced one-way random effects model with

$\sigma_i^2 \equiv \sigma^2$, the MIVQUEs for $(\sigma_\alpha^2, \sigma^2)$ obtained from (7.41) and (7.45) coincide with the ANOVA estimators in Section 2.3.1.

7.7 For the regression model, derive the expressions in (7.48) and (7.49), and the MINQUE in (7.50) for the balanced design.

7.8 For the model in (7.51) with a covariate, show that the WLS estimator for β with the *a priori* values is given by (7.53), and derive the expressions in (7.54) and (7.55) for the MINQUE.

7.9 For the model in (7.56) show that the MIVQUEs are given by (7.57) and (7.58).

CHAPTER 8

Nonnegative estimation of variance components

8.1 Introduction

As seen in the previous chapters, the ML and REML estimators of variance components are nonnegative in principle but the ANOVA and other types of estimators may take negative values. When an estimate of a variance component becomes negative, it may be replaced by a small positive quantity including zero, or the model may be changed. As alternatives to these approaches, procedures for nonnegative estimation have been proposed with their motivation derived mainly from the MINQUE principle.

As will be seen, nonnegative unbiased estimators (NNUEs) can be found for the residual variances of the ANOVA models, but not for the remaining variance components. However, such estimators can be found for certain linear combinations of the variance components. Nonnegative estimators for the individual variance components or their linear combinations can be found by ignoring the unbiasedness condition, and they may have considerably smaller MSEs than the unbiased estimators.

8.2 Invariance, unbiasedness and nonnegativeness

To estimate the variance components of the model in (6.1), consider a quadratic form $\boldsymbol{Y'AY}$ which is invariant to the fixed parameters and also nonnegative. These two conditions imply that $\boldsymbol{A}$ should be of the form $\boldsymbol{R'CC'R}$, where $\boldsymbol{R} = \boldsymbol{V_*^{-1}}[\boldsymbol{I} - \boldsymbol{X}(\boldsymbol{X'V_*^{-1}X})^- \boldsymbol{X'V_*^{-1}}]$ as defined in Section 7.3, and the matrix $\boldsymbol{C}$ is chosen suitably.

To estimate any of the σ_i^2 unbiasedly, $tr\boldsymbol{AV_i} = 1$ and $tr\boldsymbol{AV_j}{=}0$

for $(j \neq i)$. With $\boldsymbol{A}$ expressed as above, these conditions do not present any problem for estimating σ_i^2 with the properties of unbiasedness and nonnegativeness provided $\boldsymbol{V}_j$ is positive semidefinite (psd). On the other hand, consider the estimation of σ_i^2 unbiasedly when $\boldsymbol{V}_i$ is psd and $\boldsymbol{V}_j$ for $(j \neq i)$ is positive definite (pd). In this case, $tr\boldsymbol{AV}_j = 0 \Rightarrow tr\boldsymbol{R'CC'RV}_j = 0$, which implies that $\boldsymbol{A}{=}\boldsymbol{0}$. Thus, zero is the only nonnegative and unbiased estimator of σ_i^2 in this case.

In the ANOVA type of models, $\boldsymbol{V}_i$, $i = 1, 2, \ldots, p - 1$ corresponding to the random effects are positive semidefinite and $\boldsymbol{V}_p = \boldsymbol{I}$ for the residual variance σ_p^2. Hence, only σ_p^2 can be estimated unbiasedly through a nonnegative quadratic form. La Motte (1973) presents this result for nonnegative quadratic estimation. La Motte (1973a) and Pukelsheim (1981), for instance, present some conditions for the existence of such estimators. For the balanced one-way model in Section 2.3, Verdooren (1988) shows that $l_\alpha \sigma_\alpha^2 + l\sigma^2$ can have a NNUE if $l_\alpha \geq 0$ and $l \geq l_\alpha/m$.

As can be seen, for the model in (6.1), NNUE can be found for $\boldsymbol{a'\Sigma a}$ for a chosen vector $\boldsymbol{a}$, and hence for $\boldsymbol{\Sigma}$. It can also be seen that $\boldsymbol{l'\sigma}$ can have a NNUE if $\sum_i l_i \geq 0$.

It should be noted that some of the variance components of (6.1) also represent covariances between observations. For instance, in the one-way model described in Section 2.3, σ_α^2 is the variance of α_i as well as the covariance of the observations in the same group. A negative estimate in the latter case should not be perplexing.

It should also be noted that the above result on the NNUEs does not imply that invariant quadratic estimators which are biased are necessarily nonnegative. For instance, a number of linear combinations of the between and within mean squares can be considered to estimate σ_α^2 of the above model. All of them can be biased and also take negative values.

8.3 Minimum norm quadratic estimator: MINQE

P.S.R.S. Rao and Chaubey (1978) ignored the unbiasedness condition and minimized the Euclidean norm on the left hand side of (7.26); hence the name minimum norm quadratic estimator. Setting the derivative of this Euclidean norm with respect to $\boldsymbol{A}$ to zero,

$$V_* A V_* = U_* \Lambda^{1/2} \Delta \Lambda^{1/2} U_*' \ . \tag{8.1}$$

Since $AX = o$,

$$V_*^{1/2} A V_*^{1/2} = Q_* V_*^{1/2} A V_*^{1/2} Q_* \ . \tag{8.2}$$

Substituting this expression in (8.1),

$$A = R U_* \Lambda^{1/2} \Delta \Lambda^{1/2} U_*' R = \sum_i l_i (\gamma_i^4 / n_i) R V_i R \ . \tag{8.3}$$

Hence, the MINQE of σ_i^2 is given by

$$\hat{\sigma}_i^2 = (\gamma_i^4 / n_i) e' V_i e = (\gamma_i^4 / n_i) g_i \ , \tag{8.4}$$

which is nonnegative. Note that Q_* and R are the same as defined in Section 7.3.

It can be seen from (6.57) that, with $\hat{\sigma}_i^4$ replaced by γ_i^4, this estimator provides the initial step for the EM-algorithm.

8.4 Illustrations of the MINQE

8.4.1 Model with a common mean

For the model in (7.32), the MINQE for σ_i^2 takes the form of

$$\hat{\sigma}_i^2 = [(n_i - 1) s_i^2 + n_i (\overline{y}_i - \overline{y}_W)^2] / n_i \ , \tag{8.5}$$

where $W_i = n_i / \gamma_i^2$ and $\overline{y}_W = \sum_i W_i \overline{y}_i / \sum_i W_i$. If $\sigma_i^2 \equiv \sigma^2$, the MINQE for σ^2 is given by $\hat{\sigma}^2 = \sum_i \sum_j (y_{ij} - \overline{y})^2 / n$, which is the average of the squared residuals (ASR).

8.4.2 The one-way random effects model

For the unbalanced one-way random effects model described in Section 2.8, the MINQEs of σ_α^2 and σ_i^2 are given by

$$\hat{\sigma}_\alpha^2 = (\gamma_\alpha^4 / k) \sum_i W_i^2 (\overline{y}_i - \overline{y}_W)^2 \tag{8.6}$$

and

$$\hat{\sigma}_i^2 = (n_i - 1) s_i^2 / n_i + W_i^2 \gamma_i^4 (\overline{y}_i - \overline{y}_W)^2 / n_i^2 \ , \tag{8.7}$$

where $W_i = n_i / (n_i \gamma_\alpha^2 + \gamma_i^2)$ and $\overline{y}_W = \sum_i W_i \overline{y}_i / \sum_i W_i$.

If $\sigma_i^2 \equiv \sigma^2$, the estimator for σ_α^2 is obtained from (8.6) by replacing γ_i^2 with the common *a priori* value γ^2. For this case, the estimator of σ^2 is given by

$$\hat{\sigma}^2 = \frac{\sum_i (n_i - 1)s_i^2}{n} + \frac{\gamma^2}{n} \sum_i \frac{W_i^2}{n_i}(\overline{y}_i - \overline{y}_W)^2 \ , \tag{8.8}$$

where $n = \sum_i n_i$.

P.S.R.S. Rao, Kaplan and Cochran (1981) found the MSE of $\hat{\sigma}_\alpha^2$ in (8.6) to be considerably smaller than the variances of the MINQUE and ANOVA estimators of σ_α^2 and also the asymptotic variance of its ML estimator. Ahrens, Kleffé and Tenzler (1981) and Chaubey (1984) have also obtained similar results.

8.5 Minimum mean square error estimator: MIMSQE

If $\boldsymbol{Y'AY}$ is considered for estimating $\boldsymbol{l'\sigma}$, with the translation invariance and the assumption of normality for ε, following the notation of Section 7.3,

$$\begin{aligned}
MSE(\boldsymbol{Y'AY}) &= 2tr(\boldsymbol{AV_*AV_*}) + tr(\boldsymbol{AV_*} - \boldsymbol{l'\sigma})^2 \\
&= 2tr\boldsymbol{B}^2 + (tr\boldsymbol{B} - \boldsymbol{l'\sigma})^2 \ ,
\end{aligned} \tag{8.9}$$

where $\boldsymbol{V_*^{1/2}AV_*^{1/2}} = \boldsymbol{B}$.

C.R. Rao (1973) and La Motte (1973) derived the MIMSQE by minimizing this expression . From the translation invariance, $\boldsymbol{B} = \boldsymbol{Q_*BQ_*}$. Substituting this condition in (8.9) and setting its derivative with respect to $\boldsymbol{B}$ to zero,

$$2\boldsymbol{B} + (tr\boldsymbol{B} - \boldsymbol{l'\sigma})\boldsymbol{Q_*} = 0 \ . \tag{8.10}$$

From the trace of this expression,

$$tr\boldsymbol{B} = (n - q)\boldsymbol{l'\sigma}/(n - q + 2) \ , \tag{8.11}$$

where q is the rank of $\boldsymbol{X}$. Substituting (8.11) in (8.10),

$$\boldsymbol{B} = \frac{\boldsymbol{l'\sigma}}{n - q + 2}\boldsymbol{Q_*} \ . \tag{8.12}$$

Substituting the vector of the *a priori* values $\boldsymbol{\gamma}$ for $\boldsymbol{\sigma}$, from this expression,

$$\boldsymbol{A} = \frac{\boldsymbol{l'\gamma}}{n - q + 2}\boldsymbol{R} \tag{8.13}$$

and the MIMSQE of $l'\sigma$ is given by

$$l'\hat{\sigma} = \frac{l'\gamma}{n-q+2}Y'RY \; , \tag{8.14}$$

where $R = V_*^{-1}Q_*$ as defined earlier.

The MSE of this nonnegative estimator will be minimum if the *a priori* values γ_i^2 coincide with the actual variance components σ_i^2.

8.6 Proportional priors: PROPE

P.S.R.S. Rao and Chaubey (1978) noted that if $\gamma_i^2 = c\sigma_i^2$, $i = 1, 2, \ldots, p$ for a scalar c, $V_* = c\Sigma$ and $Q_*\Sigma Q_*' = Q_*\Sigma = \Sigma - X(X'\Sigma^{-1}X)^{-1}X'$ and hence

$$E[(Q_*Y)'V_i(Q_*Y)] = trV_iQ_*\Sigma Q_*' = trV_iQ_*\Sigma \; . \tag{8.15}$$

For the model in (6.1) with unequal variances, $U_i' = (0 \: \vdots \: I_i \: \vdots \: 0)$. In this case, the right hand side of (8.15) becomes $\sigma_i^2 trQ_*V_i$. Thus, an estimator for σ_i^2 is

$$\hat{\sigma}_i^2 = \frac{(Q_*Y)'V_i(Q_*Y)}{trQ_*V_i} \; . \tag{8.16}$$

This nonnegative estimator will be unbiased if $\gamma_i^2 = c\sigma_i^2$, but not otherwise.

For the replicated model,

$$y_{ij} = \beta_1 x_{i1} + \beta_2 x_{i2} + \cdots + \beta_p x_{ip} + \varepsilon_{ij} \; , \tag{8.17}$$

$i = 1, 2, \ldots, k$ and $j = 1, 2, \ldots, n_i$, Horn, Horn and Duncan (1975) derived the almost unbiased estimator (AUE). For this model, the AUE coincides with (8.16).

8.7 Spectral decomposition: SMINQUE and CMINQUE

C.R. Rao (1984) suggests the following approach. The matrix in (7.27) for the MINQUE of $l'\sigma$ can in general be expressed as

$$A = \sum_1^n a_i P_i P_i' = \sum_1^k a_i P_i P_i' - \sum_{k+1}^n a_i P_i P_i' \; , \tag{8.18}$$

where $(a_1, a_2, \ldots, a_k)$ and $(-a_{k+1} \geq -a_{k+2} \geq \ldots \geq -a_n)$ are the positive and negative eigen values of $\boldsymbol{A}$, and $\boldsymbol{P}_i$ are the corresponding eigen vectors. The MINQUE can now be expressed as

$$\boldsymbol{Y}'\boldsymbol{A}\boldsymbol{Y} = \sum_1^k a_i(\boldsymbol{P}_i'\boldsymbol{Y})^2 - \sum_{k+1}^n a_i(\boldsymbol{P}_i'\boldsymbol{Y})^2 \; . \tag{8.19}$$

If (8.19) takes a negative value, an alternative estimate that can be considered for $\boldsymbol{l}'\boldsymbol{\sigma}$ is

$$\boldsymbol{Y}'\boldsymbol{A}_1\boldsymbol{Y} = \sum_1^k a_i(\boldsymbol{P}_i'\boldsymbol{Y})^2 - \sum_{k+1}^m a_i(\boldsymbol{P}_i'\boldsymbol{Y})^2 \; , \tag{8.20}$$

where m is the largest value such that this estimate is nonnegative. This procedure can be named the SMINQUE.

To find an estimator closest to the MINQUE, the CMINQUE, Chaubey (1983) considers the minimization of $tr(\boldsymbol{A} - \boldsymbol{B}'\boldsymbol{B})^2$, where $\boldsymbol{A}$ is the matrix of the MINQUE. As a result

$$\boldsymbol{B}'\boldsymbol{B} = \sum_1^k a_i\boldsymbol{P}_i\boldsymbol{P}_i' \tag{8.21}$$

and the nonnegative estimator is given by

$$\boldsymbol{Y}'\boldsymbol{B}'\boldsymbol{B}\boldsymbol{Y} = \sum_1^k a_i(\boldsymbol{P}_i'\boldsymbol{Y})^2 \; . \tag{8.22}$$

If the MINQUE takes a negative value, (8.22) can be considered as an alternative estimator.

As can be seen, when the MINQUE takes a negative value, the SMINQUE will be closer to it than the CMINQUE and it can be very close to zero in some situations.

8.8 Convex program: CPMINQUE

Nonnegative estimates for σ_i^2 may be obtained by minimizing

$$\boldsymbol{D} = (\boldsymbol{F}\widehat{\boldsymbol{\sigma}} - \boldsymbol{g})'(\boldsymbol{F}\widehat{\boldsymbol{\sigma}} - \boldsymbol{g}) \tag{8.23}$$

with the constraints that $\hat{\sigma}_i^2 \geq 0, i = 1, 2, \ldots, p$. Hence the name CPMINQUE. The matrix $\boldsymbol{F}$ and the vector $\boldsymbol{g}$ are defined in Section 7.3. For the variances and mean of the model in (7.32), Peddada (1989) compared the CPMINQUE with other estimators.

When the sample sizes are small, the MINQE was found to have the smallest MSE followed by the CPMINQUE, SMINQE and CMINQE, in that order. The differences among these estimators were found to be negligible for large sample sizes.

8.9 Minimum bias estimator: MINBE

The bias of $Y'AY$ for estimating $l_i\sigma_i^2$ of (6.1) is $\sigma_i^2(tr\,AV_i - l_i)$. To estimate $l'\sigma$, Hartung (1981) suggests minimizing $tr\,A^2$ subject to the conditions that (1) A is positive definite and (2) $\sum_i(tr\,AV_i - p_i)^2$ is minimum.

For the one-way random effects model, Heine (1993) found that the MINBE for the residual variances σ_i^2 is the same as the sample variance $s_i^2 = \sum_j(y_{ij} - \overline{y}_i)^2/(n_i - 1)$. Comparisons for $k = 2$ and unequal n_i showed that for estimating σ_α^2, the MINQE has smaller MSE than the MINBE.

8.10 Estimators for the one-way model

The different procedures available for the nonnegative estimation of σ_α^2 are examined in this section when $\sigma_i^2 \equiv \sigma^2$ and $n_i \equiv m$. The biases and MSEs are derived with the assumption that y_{ij} follows a normal distribution. The ratio $(\sigma^2/\sigma_\alpha^2)$ is denoted by r and $(\gamma^2/\gamma_\alpha^2)$ by r_0.

8.10.1 The MINQE

From (8.6),

$$\hat{\sigma}_\alpha^2 = (w^2/k) \sum_1^k (\overline{y}_i - \overline{y})^2 \tag{8.24}$$

where $w = m/(m + r_0)$. The bias of this estimator is

$$B(\hat{\sigma}_\alpha^2) = \frac{w^2(k-1)}{k}\left(\sigma_\alpha^2 + \frac{\sigma^2}{m}\right) - \sigma_\alpha^2 , \tag{8.25}$$

which becomes negative if $r_0 > r$.

When $r_0 = 1$, this bias vanishes if

$$\rho = \frac{n - m}{3n + k + m(m-1)} < \frac{n - m}{3n + k} , \tag{8.26}$$

where $\rho = \sigma_\alpha^2/(\sigma_\alpha^2 + \sigma^2)$. The right hand side of this expression approaches $m/(3m+1)$ if k is large, $(k-1)/3k$ if m is large, and $(1/3)$ if both k and m are large.

From (8.24),

$$
\begin{aligned}
MSE(\hat{\sigma}_\alpha^2) \;=\;& \frac{2w^4}{k^2}(k-1)(\sigma_\alpha^2 + \frac{\sigma^2}{m})^2 \\
& + \left[\frac{w^2(k-1)}{k}(\sigma_\alpha^2 + \frac{\sigma^2}{m}) - \sigma_\alpha^2\right]^2 .
\end{aligned}
\tag{8.27}
$$

To obtain γ_α^2/γ^2 and hence w for this estimator, Conerly and Webster (1987) equate (8.27) to the MSE of the ML solution in (6.23) for the case of σ^2 approaching zero. Note that this MSE is given by $(2k-1)(m\sigma_\alpha^2 + \sigma^2)^2/n^2m^2$ which approaches $(2k-1)\sigma_\alpha^4/n^2$. From this procedure, $w^2 = (k-1)/(k+1)$, which implies that $\gamma_\alpha^2/\gamma^2 < (2k-1)/2m$. For γ_α^2/γ^2, the above authors suggest the larger of $(2k-1)/2m$ and unity. For unequal sample sizes in the groups, m is replaced by the harmonic mean $k/[\sum_i(1/n_i)]$.

8.10.2 The MIMSQE

From (8.14), this estimator for σ_α^2 becomes

$$
\hat{\sigma}_\alpha^2 = \frac{1}{n+1}\left[\frac{(n-k)s^2}{r_0} + w\sum_i(\bar{y}_i - \bar{y})^2\right] .
\tag{8.28}
$$

The expectation and variance of (8.28) are

$$
E(\hat{\sigma}_\alpha^2) = \frac{1}{n+1}\left[\frac{(n-k)}{r_0}\sigma^2 + w(k-1)(\sigma_\alpha^2 + \frac{\sigma^2}{m})\right]
\tag{8.29}
$$

and

$$
V(\hat{\sigma}_\alpha^2) = \frac{2}{(n+1)^2}\left[\frac{(n-k)}{r_0^2}\sigma^4 + w^2(k-1)(\sigma_\alpha^2 + \frac{\sigma^2}{m})^2\right] .
\tag{8.30}
$$

At $r_0 = r$, this estimator has a bias of $-2\sigma_\alpha^2/(n+1)$ and its MSE will attain its minimum of $2\sigma_\alpha^4/(n+1)$. At $r_0 = 1$, the bias becomes $[(nm-1)\sigma^2 - (nm+2m+1)\sigma_\alpha^2]/(n+1)(m+1)$. For large n, this bias becomes positive or negative as r is larger or smaller than unity.

8.10.3 The PROPE, MINBE and CMINQUE

From (8.16), the PROPE is given by

$$\hat{\sigma}_\alpha^2 = w \sum_i (\overline{y}_i - \overline{y})^2 / (k - 1) \; . \tag{8.31}$$

Since $E(\hat{\sigma}_\alpha^2) = w(m\sigma_\alpha^2 + \sigma^2)/m$, the bias of this estimator vanishes if $r_0 = r$; otherwise, it can be positive or negative as $r_0 < r$ or $r_0 > r$. When $r_0 = r$, $V(\hat{\sigma}_\alpha^2) = 2\sigma_\alpha^4/(k - 1)$. The MSE of this estimator attains its minimum of $2\sigma_\alpha^4/(k + 1)$ when $w = (k - 1)(m\sigma_\alpha^2 + \sigma^2)/m(k + 1)$, that is, if $r_0 > r$. For $r_0 = 1$, the bias is equal to $(\sigma^2 - \sigma_\alpha^2)/(m + 1)$.

As shown by Heine (1993), the MINBE is

$$\hat{\sigma}_\alpha^2 = \frac{nm}{nm + 1} \frac{\sum_i (\overline{y}_i - \overline{y})^2}{k - 1} \; . \tag{8.32}$$

The bias of this estimator, $(n\sigma^2 - \sigma_\alpha^2)/(nm + 1)$, becomes small as m increases.

As noted in Section 7.5.3, the MIVQUE for σ_α^2 is the same as the ANOVA estimator in (2.9). From the positive eigen values of this estimator, we find that the CMINQUE is given by

$$\hat{\sigma}_\alpha^2 = \sum_i (\overline{y}_i - \overline{y})^2 / (k - 1) \; . \tag{8.33}$$

The bias of this estimator, σ^2/m, also decreases with m.

8.10.4 An optimum estimator

As can be seen from Seely (1977), for the balanced case, $\sum_i (\overline{y}_i - \overline{y})^2$ and $\sum_i \sum_j (y_{ij} - \overline{y}_i)^2$ form a set of **complete sufficient statistics** for σ_α^2 and σ^2. Following this result, to estimate σ_α^2 we may consider $c \sum_i (\overline{y}_i - \overline{y})^2$ where c is a constant. Replacing $(\sigma_\alpha^2, \sigma^2)$ in the MSE of this estimator by $(\gamma_\alpha^2, \gamma^2)$ and minimizing it with respect to c, an optimum estimator for σ_α^2 is given by

$$\hat{\sigma}_\alpha^2 = w \sum_i (\overline{y}_i - \overline{y})^2 / (k + 1) \; . \tag{8.34}$$

At $r_0 = r$, this estimator has a negative bias equal to $2\sigma_\alpha^2/(k + 1)$ and its MSE takes the minimum value of $2\sigma_\alpha^4/(k + 1)$.

8.10.5 Further attempts

For σ_α^2, Han (1978) considers $cM_B - M_W/m$ if $F = M_B/M_W > f_\alpha$ and zero if $F \leq f_\alpha$, where f_α is the $(1-\alpha)$ percentage point of the F-distribution with $(k-1)$ and $(n-k)$ d.f., and suggests procedures for determining the constant c.

Kelly and Mathew (1993) consider $a(M_B/m - bM_W)$, where the constants $a \geq 0$ and b are found such that the MSE of this estimator is smaller than the variance of the ANOVA estimator in (2.9). They note that such an estimator can be found for only certain values of k and m, not necessarily large.

8.10.6 Comparison of the estimators

The biases and MSEs of the above six estimators are obtained by noting that the expectation and variance of $\sum_i (\overline{y}_i - \overline{y})^2$ are given by $(k-1)(m\sigma_\alpha^2 + \sigma^2)/m$ and $2(k-1)(m\sigma_\alpha^2 + \sigma^2)^2/m^2$ respectively. For the sake of illustration, the biases and the square roots of the mean square errors (RMSEs) are presented in Tables 8.1 and 8.2 for $(k, m) = (10, 5)$ and $(5, 10)$ when $\sigma_\alpha^2 = (5, 10, 20)$, $\sigma^2 = 10$ and $r_0 = 1$. The variance of the ANOVA estimator and the asymptotic variance of the ML estimators are also presented in Table 8.2. The following conclusions can be drawn from these figures.

1. As expected, the bias of the PROPE vanishes and the MSE of the MIMSQE becomes smaller than the MSEs of the remaining estimators only if $r_0 = r$.

2. If the *a priori* values differ from the actual variance components, the MINQE and OPT in general have smaller MSEs than the remaining estimators. However, these estimators will have smaller absolute biases relative to the magnitudes of σ_α^2 only if $\rho = \sigma_\alpha^2/(\sigma_\alpha^2 + \sigma^2)$ is less than fifty percent.

3. The MINQE has smaller MSE than the MINBE, although it can have a larger bias. Kleffé and J.N.K. Rao (1980) compared these estimators for large k and m.

4. The CMINQE has a bias of σ^2/m which clearly decreases with m, but this estimator has larger MSE than the MINQE and the OPT. As noted in Section 8.7, the SMINQE will be closer to the MINQUE and hence less biased than the CMINQE.

Table 8.1. *Biases of the estimators of σ_α^2 when $\sigma^2 = 10$*

σ_α^2	5	10	20	5	10	20
$\rho = \frac{\sigma_\alpha^2}{(\sigma_\alpha^2+\sigma^2)}$	1/3	1/2	2/3	1/3	1/2	2/3
	$k{=}10,\ m{=}5$			$k{=}5,\ m{=}10$		
MINQE	-0.62	-2.50	-6.25	-1.03	-2.73	-6.12
MIMSQE	3.87	-0.39	-8.92	4.25	-0.39	-9.68
PROPE	0.83	0.00	-1.67	0.46	0.00	-0.91
CMINQE	2.00	2.00	2.00	1.00	1.00	1.00
MINBE	1.97	1.95	1.91	0.99	0.98	0.96
OPT	-0.23	-1.82	-5.00	-1.36	-3.33	-7.27

Table 8.2. *RMSEs of the estimators of σ_α^2 when $\sigma^2 = 10$*

σ_α^2	5	10	20	5	10	20
$\rho = \frac{\sigma_\alpha^2}{(\sigma_\alpha^2+\sigma^2)}$	1/3	1/3	2/3	1/3	1/2	2/3
	$k{=}10,\ m{=}5$			$k{=}5,\ m{=}10$		
MINQE	2.16	4.33	9.00	2.99	5.82	11.57
MIMSQE	4.28	1.98	9.22	4.65	1.98	9.51
PROPE	2.87	4.71	8.80	3.88	7.07	13.53
CMINQE	3.30	5.66	10.37	4.24	7.78	14.85
MINBE	3.83	5.96	10.51	4.35	7.82	14.85
OPT	2.26	4.26	8.66	2.91	5.77	11.57
ANOVA	3.33	5.68	10.38	4.25	7.78	14.85
MLE*	3.13	5.37	9.84	3.80	6.96	13.28

* Square root of the asymptotic variance

Exercises

8.1 Consider the model in (6.1) for the ANOVA classifications where the residual variance σ_p^2 can have a NNUE. Let $\boldsymbol{U}_o = (\boldsymbol{U}_1 : \boldsymbol{U}_2 : \ldots : \boldsymbol{U}_{p-1})$ and note that $\boldsymbol{U}_p = \boldsymbol{I}$. Show that (a) the MINQUE for σ_p^2, $\boldsymbol{Y}'\boldsymbol{A}\boldsymbol{Y}$, which is nonnegative can be obtained by minimizing $tr\,\boldsymbol{A}^2$ with (1) $\boldsymbol{A}(\boldsymbol{X} : \boldsymbol{U}_o) = \boldsymbol{0}$ and (2) $tr\,\boldsymbol{A} = 1$, and (b) this estimator is the same as the usual residual mean squares.

8.2 As described in Section 8.2, show that NNUEs can be found for (1) $\boldsymbol{a}'\boldsymbol{\Sigma}\boldsymbol{a}$, (2) $\boldsymbol{l}'\boldsymbol{\sigma}$ if $\sum_i l_i > 0$, and (3) $l_\alpha \sigma_\alpha^2 + l\sigma^2$ if $l_\alpha \geq 0$ and $l \geq l_\alpha/m$.

8.3 Show that the bias of the MINQE vanishes if (8.26) is satisfied.

8.4 Show that the biases and the minimum MSE for the MIMSQE are as described in Section 8.10.2.

8.5 For the PROPE to attain minimum MSE, derive the conditions in Section 8.10.3.

8.6 Find the eigen values and vectors of the MINQUE for σ_α^2 of the one-way balanced model, and show that the CMINQUE is given by (8.33).

8.7 Derive the estimator in (8.34) and its minimum MSE at $r_0 = r$.

CHAPTER 9

Confidence intervals

9.1 Introduction

As seen in Chapter 2 for the balanced one-way classification, exact confidence limits can be found for a ratio of the variance components, but this problem can become complicated for unbalanced designs.

The distributions of the estimators of individual variance components and their linear combinations usually do not have simple forms even for balanced designs and will not be suitable for finding the confidence limits. For instance, the distributions of $\hat{\sigma}_\alpha^2$ in (2.9) for balanced designs and (2.35) for unbalanced designs do not have simple forms for finding the confidence limits for σ_α^2.

Approximate procedures have been suggested for the confidence limits of individual variance components and their linear combinations. Some of these procedures can have desirable widths and coverage probabilities

9.2 Chi-square approximation

Denoting the variance components by $\theta_i, i = 1, 2, \ldots, k$, Satterthwaite's (1946) approximation in Section 1.3.4 can be used for finding the confidence limits for a linear combination $\theta = \sum_i a_i \theta_i$, where a_i are chosen constants. The unbiased estimators of θ_i are denoted by M_i and of θ by M. The total $T = \sigma_\alpha^2 + \sigma^2$ and the variance of the mean $V(\bar{y}) = \sigma_\alpha^2 + \sigma^2/m$ of the balanced one-way model in Section 2.3 are two illustrations of the above linear combination.

From (2.8), an unbiased estimator for T is

$$\hat{T} = \frac{\sum_i(\bar{y}_i - \bar{y})^2}{k - 1} + \frac{m - 1}{m}s^2 = \frac{M_B + (m - 1)M_W}{m} \tag{9.1}$$

With the notation of Section 1.3.4, $a_1 = 1, a_2 = (m - 1)/m$, $M_1 = \sum_i(\bar{y}_i - \bar{y})^2/(k - 1) = M_B/m$, and $M_2 = s^2 = M_W$. The d.f. for M_1 and M_2 are $f_1 = (k - 1)$ and $f_2 = (n - k)$. Following the above approach, $f\hat{T}/T$ is approximated by a χ^2-distribution with d.f.

$$\begin{aligned} f &= \frac{[M_1 + (m - 1)M_2/m]^2}{M_1^2/(k - 1) + (m - 1)^2 M_2^2/m^2(n - k)} \\ &= \frac{[M_B + (m - 1)M_W]^2}{M_B^2/(k - 1) + (m - 1)^2 M_W^2/(n - k)} \end{aligned} \tag{9.2}$$

Confidence limits for T are obtained from $f\hat{T}/\chi_l^2$ and $f\hat{T}\chi_u^2$, where χ_l^2 and χ_u^2 are the lower and upper percentage points of the χ^2-distribution with f d.f.

Example 9.1 Ozone depletion

For the balanced model described in Section 2.3, and the ozone data in Table 2.1, $M_B = 9.09$ and $M_W = 0.9256$. Substituting these figures in (9.2), $f = 4.97$ or approximately 5. The estimate of T is $\hat{T} = 2.1$. When $(1 - \alpha) = 0.95$, the lower and upper percentage points of the χ^2-distribution with 5 d.f. are (0.83, 12.38). With these values, confidence limits for T are given by $5(2.1)/12.38 = 0.85$ and $5(2.1)/(0.83) = 12.65$.

9.3 Normal approximation

When M_i have independent χ^2-distributions with f_i d.f., the variance of M is $2\sum_i a_i^2\theta_i^2/f_i$, which may be estimated from $2\sum_i a_i^2 M_i^2/f_i$. If M is assumed to have a normal distribution for large sample sizes, confidence limits for θ can be obtained from

$$M - Z\sqrt{2\sum_i a_i^2 M_i^2/f_i} \quad , \quad M + Z\sqrt{2\sum_i a_i^2 M_i^2/f_i} \, , \tag{9.3}$$

where Z is the standard normal deviate for a specified probability. In this expression, M_i^2 may be replaced by the unbiased estimator

$f_i M_i^2/(f_i+2)$ of θ_i^2. Welch (1956) suggests this procedure of finding the confidence limits and further refinements.

Example 9.2 Ozone depletion

For the illustration in the last section, $a_1 = 1$, $a_2 = 6/7$, $M_1 = 9.09/7 = 1.2986$ and $M_2 = 0.9256$. Hence, an estimate of $V(\hat{T})$ is 0.91. Ninety five percent confidence limits for T are given by $2.1 - 1.96(0.95) = 0.24$ and $2.1 + 1.96(0.95) = 3.96$.

Bulmer (1957) and Huitson (1955) also suggest approximate confidence limits for linear combinations of variance components. For a linear combination of the one-way unbalanced model, $a_1\sigma_\alpha^2 + a_2\sigma^2$, where a_1 and a_2 are positive constants, Burdick and Siellken (1978) developed exact limits. Seely and Lee (1994) modify Satterthwaite's procedure to find confidence limits for σ_α^2.

9.4 Modified large sample (MLS) approximation

Graybill and Wang (1980) consider a modification to the above normal approximation. For the sake of illustration, consider $\theta = a_1\theta_1 + a_2\theta_2$. The $(1-\alpha)$ percent interval for θ can be expressed as

$$
\begin{aligned}
M \quad &- \quad (L_1^2 a_1^2 M_1^2 + L_2^2 a_2^2 M_2^2)^{1/2} \le \theta \\
&\le \quad M + (H_1^2 a_1^2 M_1^2 + H_2^2 a_2^2 M_2^2)^{1/2} ,
\end{aligned} \tag{9.4}
$$

The constants (L_1, L_2, H_1, H_2) are determined by considering $\theta_1 = 0$ and $\theta_2 = 0$ separately.

First, $\theta_1 = 0$ implies that $M_1 = 0$, and in this case the probability corresponding to (9.4) can be expressed as

$$
\begin{aligned}
P[(1 - L_2)M_2 \le \quad &\theta_2 \quad \le (1 + H_2)M_2] \\
&= \quad P[\frac{f_2}{1 + H_2} \le \frac{f_2 M_2}{\theta_2} \le \frac{f_2}{1 - L_2}] \\
&= \quad (1 - \alpha) .
\end{aligned} \tag{9.5}
$$

Since $f_2 M_2/\theta_2$ has a χ^2-distribution with f_2 d.f., $f_2/(1 + H_2)$ and $f_2/(1-L_2)$ are respectively equal to the lower and upper percentage points $\chi^2(\alpha_{21}, f_2)$ and $\chi^2(1 - \alpha_{22}, f_2)$, where $(\alpha_{21} + \alpha_{22}) = \alpha$. Thus, $L_2 = 1 - [f_2/\chi^2(\alpha_{21}, f_2)]$ and $H_2 = [f_2/\chi^2(1 - \alpha_{22}, f_2)] - 1$. Similarly, with $\theta_2 = 0$, $L_1 = 1 - [f_1/\chi^2(\alpha_{11}, f_1)]$ and $H_1 = [f_1/\chi^2(1 - \alpha_{12}, f_1)] - 1$, where $(\alpha_{11} + \alpha_{12}) = \alpha$.

Burdick and Graybill (1992) present further extensions of this approach and provide the related bibliography.

9.5 An approximation with the USS estimator

For the one-way unbalanced model discussed in Section 2.7, $S_B = \sum_i n_i(\overline{y}_i - \overline{y})^2$ does not follow a χ^2-distribution unless $\sigma_\alpha^2 = 0$. As an alternative, with the USS estimator for σ_α^2 in (2.39), an unbiased estimator for T is given by

$$\hat{T} = \frac{\sum_i(\overline{y}_i - \overline{y}_U)^2}{k-1} + \frac{H-1}{H}s^2 \, . \tag{9.6}$$

In this expression, $\overline{y}_U = \sum_i \overline{y}_i/k$ and $H = k/\sum_i n_i^{-1}$ is the harmonic mean of the sample sizes. The expectation of $\sum_i(\overline{y}_i - \overline{y}_U)^2$ is equal to $\sigma_\alpha^2 + \sigma^2/H$. Thomas and Hultquist (1978) show that $\sum_i(\overline{y}_i - \overline{y}_U)^2/(\sigma_\alpha^2 + \sigma^2/H)$ approximately follows a χ^2-distribution with $(k-1)$ d.f. As can be seen from this expression, the approximation becomes valid as σ_α^2/σ^2 is large or n_i are large. From these results, $f\hat{T}/T$ approximately follows a χ^2-distribution where the degrees of freedom f are obtained from (9.2) by replacing m by H, M_B by $H\sum_i(\overline{y}_i - \overline{y}_U)^2/(k-1)$ and M_1 by M_B/H. Confidence limits for T are obtained as before from $f T/\chi_l^2$ and $f\hat{T}/\chi_u^2$. Burdick and Eickman (1986) suggest a modification to this procedure.

Example 9.3 Ozone depletion

For the unbalanced model in Section 2.7 and the ozone data in Table 2.3, $H = 4.94$, $\overline{y}_U = 2.33$, $\sum_i(\overline{y}_i - \overline{y}_U)^2 = 3.96$, $s^2 = 1.04$, $\hat{T} = 2.15$ and $f = 7.44$ or approximately 7. For $(1-\alpha) = 0.95$, $\chi_l^2 = 1.69$ and $\chi_u^2 = 16$. From these figures confidence limits for T are $7(2.15)/16 = 0.94$ and $7(2.15)/1.69 = 8.91$.

9.6 A single variance component

For the variance component σ_α^2 of the one-way balanced model in Section 2.3.1, Tukey (1951) and Williams (1962) independently developed confidence limits from the interval common to $(\sigma_\alpha^2/\sigma^2)$ and $(m\sigma_\alpha^2 + \sigma^2)$.

From (2.16), $(1-\alpha)$ percent confidence interval for σ_α^2 for a given σ^2 is

$$\left(\frac{F}{F_u}-1\right)\frac{\sigma^2}{m} \leq \sigma_\alpha^2 \leq \left(\frac{F}{F_l}-1\right)\frac{\sigma^2}{m} \tag{9.7}$$

Since $S_B/(m\sigma_\alpha^2+\sigma^2)$ follows a χ^2-distribution with $(k-1)$ d.f., $(1-\alpha)$ percent confidence interval for $(m\sigma_\alpha^2+\sigma^2)$ is given by

$$(S_B/\chi_u^2) \leq (m\sigma_\alpha^2+\sigma^2) \leq (S_B/\chi_l^2) . \tag{9.8}$$

In this expression, (χ_l^2,χ_u^2) are the lower and upper $(1-\alpha)$ percentage points of the χ^2-distribution with $(k-1)$ d.f. For a given σ^2, (9.8) can be expressed as

$$\frac{S_B}{m\chi_u^2}-\frac{\sigma^2}{m} \leq \sigma_\alpha^2 \leq \frac{S_B}{m\chi_l^2}-\frac{\sigma^2}{m} . \tag{9.9}$$

Denoting the intervals in (9.7) and (9.9) by I_1 and I_2, from the Bonferroni type of inequality, $P(I_1 \cap I_2) \geq 1-2\alpha$. The lower limit for the common interval is obtained by finding σ^2 from equating the left hand sides of (9.7) and (9.9). Similarly, the upper limit is obtained from the right hand sides. From this procedure,

$$\left(1-\frac{F_u}{F}\right)\frac{S_B}{m\chi_u^2} \leq \sigma_\alpha^2 \leq \left(1-\frac{F_l}{F}\right)\frac{S_B}{m\chi_l^2} \tag{9.10}$$

and the probability of this interval enclosing σ_α^2 is at least $(1-2\alpha)$.

Moriguti (1954) developed approximate confidence limits for σ_α^2 from the distributions of $S_B/(m\sigma_\alpha^2+\sigma^2)$ and S_W/σ^2. Boardman (1974) conducted a simulation study to examine the different procedures for the balanced one-way model. The coverage probabilities for the procedures of Tukey (1951), Williams (1962), Moriguti (1954) and Bulmer (1957) were found to be satisfactory in general; Satterthwaite's approximation performed well for $\sigma_\alpha^2/\sigma^2 \geq 3$. Bross (1946) and Venables and James (1978) presented the fiducial limits for σ_α^2.

The following procedures were also suggested to obtain confidence limits for σ_α^2, for instance, by Anderson and Bancroft (1952): (a) Although Satterthwaite's approximation for the distribution of $\sum_i a_i M_i$ is not valid when some of the a_i are negative, it may be attempted for the distribution of $\hat{\sigma}_\alpha^2$ in (2.9); (b) With the estimate of the variance of $\hat{\sigma}_\alpha^2$ in (2.11), the normal approximation may be employed; (c) Confidence limits for σ_α^2 may also be found by replacing σ^2 in (9.7) or (9.8) by its estimate s^2. If σ_α^2 is much larger than σ^2 and n_i are large, these procedures may have the desirable confidence widths and coverage probabilities. Sahai (1974) and Khuri (1981) develop simultaneous confidence intervals for variance components.

9.7 Related topics

9.7.1 Ratios of linear combinations

Cochran (1951) suggests tests of hypotheses for the variance components of mixed models. These tests are expressed as ratios of linear combinations of mean squares and they can be assumed to follow the F-distributions with the d.f. determined from Satterthwaite's approximation.

Howe and Myers (1970) and Myers and Howe (1971) suggest alternative procedures for finding the degrees of freedom for the above type of approximation to the F-distribution. Davenport (1975) and Davenport and Webster (1973) examine the validity of these procedures.

9.7.2 Ratios of variance components

Exact confidence limits for σ_α^2/σ^2 of the one-way balanced model were presented in (2.16). Confidence limits for similar ratios of variances or expected mean squares for other models and unbalanced designs have appeared in the literature; see for instance, Bromerling (1969), Verdooren (1988), Lu, Graybill and Burdick (1989), and Wang and Graybill (1991).

Exercises

9.1 For the balanced case in Example 9.1, find the estimates and ninetyfive percent confidence limits for T if M_B and M_W remain the same as 9.09 and 0.9256 when (1) $k = 2$, $m = 10$ and (2) $k = 4$ and $m = 5$; use the approximation in (9.2). From the results of the above example and these two cases, examine the effects of k and m on the confidence limits.

9.2 For the balanced case in Examples 9.1 and 9.2, find ninetyfive percent confidence limits for T from the MLS approximation in (9.4) and compare them with the limits in the above two examples.

9.3 For the unbalanced design in Section 2.7 and the data in Table 2.3, estimate T from M_B and M_W. Assuming that M_B/m_o follows a χ^2-distribution with $(k-1)$ d.f. independent of M_W, find 95 percent confidence limits for T and compare them with the limits obtained in Example 9.3.

9.4 From the data in Table 2.1, find 95 percent confidence limits for σ_α^2 of the balanced model in Section 2.3.1 through (9.10) and the three approximate procedures described in Section 9.6.

9.5 Describe how the sum of squares $\sum_i (\bar{y}_i - \bar{y}_U)^2$ in (9.6) can have a χ^2-distribution for large $(\sigma_\alpha^2/\sigma^2)$ or large n_i.

9.6 From the ozone data in Table 2.1, find 95 percent confidence limits for the intraclass correlation coefficient.

CHAPTER 10

Combining information from experiments

10.1 Introduction

For increased applicability, many agricultural, biological and medical experiments are frequently repeated a few times at the same location, or at more than one location or center during the same period of time. Estimates for the means or treatment effects along with their standard errors usually become available from all the experiments. Three illustrations of such replicated experiments are presented in the following section.

Cochran (1937, 1938, 1954) and Yates and Cochran (1938) examined the maximum likelihood and weighted least squares methods for combining the estimates and testing the related hypothesis. These procedures reviewed by P.S.R.S. Rao (1984) are applicable for combining a small number of studies or the meta-analysis of a large number of studies. The combined estimates in general will have smaller standard errors than the individual estimates. Adjusting the estimates for concomitant or supplementary information can have additional advantages.

10.2 Illustrations

(a) Weight reduction

From a study on the effect of a diet for weight reduction, conducted at k locations on n_i candidates at the ith location, $i = 1, 2, \ldots, k$, the average loss in weight y_i and its standard error s_i may become available. Denoting the variance per observation with $\gamma_i = (n_i - 1)$

d.f. by s_{oi}^2, $s_i = s_{oi}/\sqrt{(n_i)}$.

(b) Medical treatments

From the responses to a medical treatment and a control at each of k participating centers, the differences of their averages y_i for $i = 1, 2, \ldots, k$ and the standard errors $s_i = s_{oi}/\sqrt{d_i}$ of y_i become available. In this case, s_{oi}^2 is the pooled variance with $\nu_i = (n_i + m_i - 2)$ d.f., where n_i and m_i are the sizes of the treatment and control groups, and $d_i = n_i m_i/(n_i + m_i)$. Similar figures are usually reported for specified contrasts of the treatment effects, for instance, from a one-way or two-way classification.

(c) Randomized blocks

Consider a randomized block experiment with b blocks and t treatments replicated at k locations. For the difference y_i of two treatment means, the standard error takes the form of $s_i = s_{oi}/\sqrt{d}$, where s_{oi}^2 is the variance per observation with $\nu = (b-1)(t-1)$ d.f. and $d = b/2$. If this experiment is repeated at the same location, the variance per observation is given by $s_o^2 = \sum_i s_{oi}^2/k$ with $k\nu$ d.f. and the S.E. of y_i in this case becomes $s_o/\sqrt{d}$.

10.3 A model and the WLS estimator

Following Cochran (1937), the estimators y_i at the k centers or locations can be represented by the model

$$y_i = \mu + \alpha_i + \varepsilon_i \,, \tag{10.1}$$

$i = 1, 2, \ldots, k$. The random errors ε_i are assumed to have independent normal distributions, with zero means and variances σ_i^2. These variances can also be expressed as σ_{oi}^2/d_i, where σ_{oi}^2 is the variance per observation and d_i depends on the particular linear combination considered for y_i. The random effect α_i is assumed to have a normal distribution, independent of ε_i, with zero mean and variance σ_α^2. The variance σ_i^2 represents the internal or local variability of the estimator at the ith location, and σ_α^2 represents variability among the k estimators.

The WLS estimator for μ is

$$\overline{y}_W = \sum_i W_i y_i / W , \qquad (10.2)$$

where $W_i = 1/(\sigma_\alpha^2 + \sigma_i^2)$ and $W = \sum_i W_i$. This estimator is unbiased for μ and has variance $V(\hat{\mu}) = 1/W$. The weights W_i can be estimated from y_i and s_i as follows.

The unweighted mean of the estimates is $\overline{y} = \sum_i y_i / k$; the subscript (U) is suppressed throughout this chapter. The between mean square $s_b^2 = \sum_i (y_i - \overline{y})^2 / (k-1)$ is unbiased for $\sigma_\alpha^2 + \sum_i \sigma_i^2 / k$ and s_i^2 for σ_i^2. Hence an unbiased estimator for σ_α^2 is given by

$$\hat{\sigma}_\alpha^2 = s_b^2 - s^2 , \qquad (10.3)$$

where $s^2 = (\sum_i s_i^2)/k$.

The estimator for μ now is given by

$$\overline{y}_w = \sum_i w_i y_i / w , \qquad (10.4)$$

where $w_i = 1/(\hat{\sigma}_\alpha^2 + s_i^2)$ and $w = \sum_i w_i$. If k and the degrees of freedom ν_i for $s_i^2, i = 1, 2, \ldots, k$, are large, the variance of this estimator approximately becomes $1/W$. Estimating this variance by $1/w$, a test for $\mu = 0$ is given by $\sqrt{w}\,\overline{y}_w$, which can be approximated by the normal distribution.

Example 10.1 Effect of salt reduction on blood pressure

Physicians routinely advise their hypertensive patients to reduce the dietary intake of salt, and the effects of such a practice continues to be studied at several clinical centers. Some of these studies are conducted through cross-over designs in which treatments are administered to patients in alternating sequences over intervals of time. Average reductions of systolic and diastolic blood pressures (BPs) and their standard errors for six such studies obtained from the summary figures of Cutler *et al.* (1991) are presented in Table 10.1. The duration of these studies ranged from four to six weeks.

For the systolic blood pressure, $s^2 = 9.53$, $s_b^2 = 18.22$ and hence $\hat{\sigma}_\alpha^2 = 8.69$. From these results, $\overline{y}_w = 4.69$ and S.E.$(\overline{y}_w) = 1.64$. Ninetyfive percent confidence limits for μ are $(1.48, 7.91)$.

Table 10.1. *Decrease in blood pressure (mmHg) following reduced intake of salt*

| | | Systolic BP | | Diastolic BP | |
No.	ν_i	y_i	s_i	y_i	s_i
1	18	10.0	3.06	5.0	1.78
2	17	0.5	1.50	0.3	0.64
3	11	5.2	4.10	1.8	3.55
4	39	0.8	1.80	0.8	1.55
5	19	8.0	2.60	5.0	1.62
6	8	9.7	4.33	5.1	2.94

Source: Cutler *et al.* (1991)

The estimates in Table 10.1 range from 0.5 to 10 indicating that possibly not all of them were obtained from the same type of designs or they were affected by some factors related to blood pressures. Yates and Cochran (1935) emphasize that only estimates obtained from similar experiments should be combined. Cochran (1954) cautions against combining estimates that vary widely.

In the above type of medical experiments, the treatment effects are usually adjusted for covariates such as initial BP, age and weight of the participating candidates. It should be noted that for combining the effects of a treatment from the experiments, all of them should be adjusted through the same set of covariates.

For the meta-analysis of clinical trials, DerSimonian and Laird (1986) consider the model in (10.1) with suitable modifications, Begg and Pilote (1991) consider a model that incorporates controls, and Stram (1996) suggests a mixed model.

10.4 A test of hypothesis for σ_α^2

The estimator in (10.2) depends on the magnitudes of σ_α^2 and σ_i^2. A test for $\sigma_\alpha^2 = 0$ is given by s_b^2/s^2 which can be approximated by the F-distribution $F(f_1, f_2)$. From (1.5), the d.f. are given by

$$f_1 = (k-1)^2 s^4 / [(k-2) \sum_i s_i^2 / k + s^4]$$

and

$$f_2 = (\sum_i s_i^2)^2 / \sum_i (s_i^4 / \nu_i) . \tag{10.5}$$

For the systolic blood pressures in Table 10.1, $s_b^2 / s^2 = 1.91$, $f_1 = 4$ and $f_2 = 42$. Since $F(4, 42; 0.05) = 2.6$, it can be inferred that σ_α^2 is close to zero.

10.5 Inference regarding the mean when $\sigma_\alpha^2 = 0$

If it can be assumed through the F-test in Section 10.4 or otherwise that $\sigma_\alpha^2 = 0$, (10.2) becomes $\sum_i (y_i / \sigma_i^2) / \sum_i (1/\sigma_i^2)$ which can be estimated from $\overline{y}_w = \sum_i w_i y_i / w$, where $w_i = 1/s_i^2$ and $w = \sum_i w_i$.

Since the expectation of this estimator conditional on s_i^2, $i = 1, 2, \ldots, k$, is unbiased for μ, it is unconditionally unbiased. The conditional variance of this estimator is $\Sigma w_i^2 \sigma_i^2 / w^2$, which may be estimated from $1/w$. For large k and ν_i, this estimator may also be considered for the unconditional variance $V(\overline{y}_w)$. Cochran (1937), Meier (1953) and Cochran and Carroll (1954) present approximations to $V(\overline{y}_w)$ and its estimator.

Note that if $f(1/w)/E(1/w)$ is assumed to have a χ^2-distribution with f d.f., from (1.5), $f = w^2 / \sum_i (w_i^2 / \nu_i)$. A test for $\mu = 0$ is provided by $\sqrt{w}\, \overline{y}_w$ which approximately follows the t-distribution with f d.f.

Example 10.2 Effect of salt reduction on blood pressure (cont.)

If it is assumed from the test in Section 10.4 or otherwise that $\sigma_\alpha^2 = 0$, from (10.6) $\overline{y}_w = 3.55/1.12 = 3.17$. An estimate of the variance of $\overline{y}_w$ is $1/1.12 = 0.89$ and hence S.E.$(\overline{y}_w) = 0.94$.

From these figures, $t = 3.17/0.94 = 3.37$ with approximate d.f. of $f = (1.12)^2 / 0.0165 = 76$. This value of t is significant even at the 0.001 level. Approximate ninetyfive percent confidence limits for μ are $(1.29, 5.05)$.

From the results of this and the previous example, we find that the reduction of dietary salt can reduce systolic blood pressure.

If σ_α^2 is found to be significantly larger than $\sigma_i^2, i = 1, 2, \ldots, k$, μ can be estimated from the unweighted mean $\bar{y}$. The variance of this estimator is $V(\bar{y}) = \sum_i (\sigma_\alpha^2 + \sigma_i^2)/k^2$ which can be unbiasedly estimated from $v(\bar{y}) = s_b^2/k$.

10.6 Equal standard errors

If all the k experiments are repeated at the same location, σ_{oi}^2 for $i = 1, 2, \ldots, k$ can be assumed to be equal to σ_o^2. In this case, the variance of a specified linear combination y_i becomes $\sigma_i^2 \equiv \sigma^2 \equiv \sigma_o^2/d$. As described in Section 10.2 for the randomized block experiment repeated at the same location, the variance of the difference of two treatment effects takes this form.

The estimator for μ in this case is the same as the unweighted mean $\bar{y}$, which is unbiased and has the variance $V(\bar{y}) = (\sigma_\alpha^2 + \sigma^2)/k$. An unbiased estimator of this variance is $v_1(\bar{y}) = s_b^2/k$.

From the sample variance s_i^2 with ν d.f., the pooled variance $s^2 = \sum_i s_i^2/k$ with νk d.f. is unbiased for σ^2. A test for σ_α^2 is provided by $F = s_b^2/s^2$ which follows the F-distribution with $(k-1)$ and νk d.f.

The estimator $v_1(\bar{y})$ is also unbiased for $V(\bar{y})$ even when $\sigma_\alpha^2 = 0$. However, an alternative unbiased estimator in this case is $v_2(\bar{y}) = s^2/k$. Combining both these estimators, the minimum variance unbiased estimator for $V(\bar{y})$ is

$$v_3(\bar{y}) = \frac{\nu k s^2 + (k-1)s_b^2}{k(\nu k + k - 1)} . \tag{10.6}$$

Following the studies of Bancroft (1944) and Paull (1950), Cochran (1954) investigates the conditions suitable for the three estimators. He recommends $v_3(\bar{y})$ only if $F < 2$, and notes that the precision of this estimator is not much greater than that of $v_2(\bar{y})$ unless ν and k are small.

When σ_i^2 takes the form of σ_o^2/d_i and ν_i are different, the estimators in the previous sections can be suitably modified. In some cases σ_{oi}^2 is assumed to be the same for all the experiments and s_{oi}^2 may become available. For all these different situations, Cochran (1954) discusses in detail the estimation of μ and the standard error.

10.7 Partial weighting

For the case of $\sigma_\alpha^2 = 0$, if $g(< k)$ of the s_i^2 are much smaller than the remaining variances, the resulting w_i will have an unduly large influence on $\overline{y}_w$. When the g variances are smaller than a specified value s_o^2, Cochran (1937) considers replacing the corresponding w_i by $1/s_o^2$; the remaining $(k - g)$ weights retain their original values $1/s_i^2$. Since it is difficult to choose s_o^2, Yates and Cochran (1938) replace the g weights by their average.

For the sake of robustness against nonnormality of y_i and as a protection against small values of s_i^2, Mosteller and Tukey (1982, 1983) rank the s_i^2 and suggest procedures for finding the weights for the estimator in (10.2) and the standard error of the resulting estimator. Halvorsen (1984) compares these procedures through empirical studies and also considers them with the MINQE type of estimators for σ_i^2.

10.8 Maximum likelihood estimator

For the likelihood of $(\mu, \sigma_\alpha^2, \sigma_i^2)$, from the joint distribution of y_i and s_i^2 we find that except for a constant

$$
\begin{aligned}
-2lnL \;=\; & \sum_i ln(\sigma_\alpha^2 + \sigma_i^2) + \sum_i \nu_i ln(\sigma_i^2) \\
+\; & \sum_i [(y_i - \mu)^2/(\sigma_\alpha^2 + \sigma_i^2)] + \sum_i (\nu_i s_i^2/\sigma_i^2) \,. \quad (10.7)
\end{aligned}
$$

When $\nu_i \equiv \nu$, Cochran (1937) examines the ML estimator for μ for the cases of $\sigma_\alpha^2 > 0$ and $\sigma_\alpha^2 = 0$. From his investigations it was found that if σ_i^2 vary the ML estimator has larger precision than $\overline{y}_W$ or $\overline{y}_w$ provided ν is small; for instance, 20 or less. If $\nu > 20$, these WLS estimators can be preferred. The unweighted mean $\overline{y}$ will have higher precision than both the ML and WLS estimators when σ_i^2 do not vary much or if σ_α^2 is large relative to σ_i^2.

10.9 Supplementary information

Along with the estimates for a linear combination of the parameters or treatment combinations of interest, frequently supplementary information related to them becomes available from the k studies or experiments. For studies on blood pressures and cholesterol levels, this additional information is provided by ages, weights and similar characteristics of the participating candidates. The overall estimates can be obtained by adjusting the individual estimates for the additional information.

With a single supplementary variable x_i, a model that can be considered is

$$
\begin{aligned}
y_i &= \delta_i + \beta x_i + \varepsilon_i \\
    &= \delta + \beta x_i + \alpha_i + \varepsilon_i \ ,
\end{aligned}
\tag{10.8}
$$

$i = 1, 2, \ldots, k$, where $\delta_i = \delta + \alpha_i$. The assumption for the random effect α_i and the residual ε_i are the same as in Section 10.3. This model with its assumption for δ_i represents the situation where the slope β is the same for all the k studies, but the intercepts are random with mean δ and variance σ_α^2.

With the above assumptions, the WLS estimators for β and δ are

$$
\hat{\beta} = \sum_i W_i(x_i - \overline{x}_W)y_i / \sum_i W_i(x_i - \overline{x}_W)^2
$$

and

$$
\hat{\delta} = \overline{y}_W - \hat{\beta}\overline{x}_W \ ,
\tag{10.9}
$$

where $W_i = 1/(\sigma_\alpha^2 + \sigma_i^2)$ and $\overline{y}_W = \sum_i W_i y_i / W$ as defined before and $\overline{x}_W = \Sigma W_i x_i / W$. The estimator for $E(y_i | x_i)$ is

$$
\hat{y}_i = \overline{y}_W + \hat{\beta}(x_i - \overline{x}_W) \ .
\tag{10.10}
$$

The variance of this estimator is given by

$$
V(\hat{y}_i) = \frac{1}{W} + \frac{(x_i - \overline{x}_W)^2}{\Sigma W_i(x_i - \overline{x}_W)^2} \ .
\tag{10.11}
$$

The estimator and the variance in (10.10) and (10.11) depend on σ_α^2 and σ_i^2. Since s_i^2 is unbiased for σ_i^2, σ_α^2 can be estimated as follows. With the unweighted means $\overline{x} = \sum_i x_i / k$ and $\overline{y} = \sum_i y_i / k$ and the estimator for the slope $b = \sum_i (x_i - \overline{x})y_i / \sum_i (x_i - \overline{x})^2$, the residuals are given by $e_i = (y_i - \overline{y}) - b(x_i - \overline{x})$. Now, from (10.8),

$$
E(\sum_i e_i^2) = (k - 2)\sigma_\alpha^2 + (k - 1)(\sum_i \sigma_i^2)/k - \sum_i c_i \sigma_i^2 \ ,
\tag{10.12}
$$

Table 10.2. *Decrease in blood pressure and weight loss*

No.	ν	Systolic y_i	Systolic s_i	Diastolic y_i	Diastolic s_i	Weight loss x_i
1	24	45	6	35	5	30
2	19	45	4	30	4	15
3	34	40	4	25	3	15
4	23	35	5	20	2	10
5	14	20	5	15	3	4
6	12	20	4	5	2	2

where $c_i = (x_i - \overline{x})/\sum_i (x_i - \overline{x})^2$. From this expression,

$$\hat{\sigma}_\alpha^2 = [\sum_i e_i^2 - (k-1)s^2 + \sum_i c_i s_i^2]/(k-2) , \qquad (10.13)$$

where $s^2 = \sum_i s_i^2/k$.

Note that σ_α^2 and σ_i^2 and the estimator in (10.10) can also be obtained from the REML or MIVQUE methods through iterative procedures.

Example 10.3 Blood pressure treatment and weight loss

Hypertensive patients are advised to reduce their weights as they are treated with medication. Averages of the decreases in blood pressures and their S.E.s along with the averages of the weight losses after a specified period of time are presented in Table 10.2 for six studies. These figures are constructed to examine the decrease in blood pressure (y_i) after adjusting for weight loss (x_i).

For the systolic blood pressure, $\overline{y} = 34.2$, $s^2 = 22.33$, $s_b^2 = 134.17$ and hence $\hat{\sigma}_\alpha^2 = 111.84$. The F-ratio is $(134.17/22.33) = 6.01$, with $f_1 = 5$ and $f_2 = 107$ d.f., which is significant at the 0.01 level. An estimate for μ with $w_i = 1/(\hat{\sigma}_\alpha^2 + s_i^2)$ is $\overline{y}_w = 34.06$ which has a S.E. of 4.72. Approximate 95 percent confidence limits for μ are (24.81, 43.31).

With the data on weight loss, $b = 0.9823$ and from (10.13), $\hat{\sigma}_\alpha^2 =$

10.09. Now $\bar{y}_w = (6.5677/0.1937) = 33.91$, $\bar{x}_w = (2.2762/0.1937)$ $= 11.75$. With these figures, from (10.9), $\hat{\beta} = (14.7461/13.4776) = 1.0941$ and $\hat{\delta} = 21.0543$.

It is of interest to estimate the reduction in the systolic blood pressure, for instance, at $x_i = 0, 10$ and 30. (a) At $x_i = 0$, $\hat{y}_i = \hat{\delta} = 21.05$ and from (10.11) S.E.$(\hat{y}_i) = 3.93$. With the normal approximation for $\hat{y}_i$, ninety five percent confidence limits for $E(y_i|x_i = 0)$ are $(13.35, 28.75)$. (b) At $x_i = 10$, $\hat{y}_i = 32.0$, S.E.$(\hat{y}_i) = 2.32$ and the confidence limits are $(27.45, 36.55)$. (c) At $x_i = 30$, $\hat{y}_i = 53.88$, S.E.$(\hat{y}_i) = 5.47$ and the confidence limits are $(43.16, 64.60)$.

Exercises

10.1 Find the estimate and 95 percent confidence limits for the overall mean of the systolic blood pressure from excluding the figures for the second and fourth study in Table 10.1, and compare them with the results of Example 10.1.

10.2 From the data in Table 10.1, find the overall estimate and 95 percent confidence limits for the mean diastolic pressure.

10.3 Adjust the data on the diastolic blood pressures in Table 10.2 for the weight loss and find the estimate, standard error and 95 percent confidence limits for the mean.

10.4 Haines, Komov and Jagoe (1992) examined the acidity and mercury content of perch, a type of small fish, through samples collected from the Rybinsk reservoir and six lakes. The reservoir and lakes are situated adjacent to the Darwin National Reserve, a 'protected natural area.' From samples of size nine from the reservoir, five from the first lake and ten from each of the remaining five lakes, the averages and standard deviations of the mercury content in micrograms per gram of weight of perch for the reservoir followed by the lakes were found to be $(0.19, 0.03)$, $(0.78, 0.05)$, $(0.50, 0.02)$, $(0.64, 0.01)$, $(1.06, 0.22)$, $(0.57, 0.03)$ and $(0.11, 0.01)$. (a) Find the estimate and 95 percent confidence limits for the overall mean of the mercury content and (b) test for the significance of the variance component.

10.5 For the case of $\sigma_\alpha^2 = 0$ discussed in Section 10.5, find the exact expectations of w_i and w_i^2 and an approximation to the expectation of the conditional variance $\sum_i w_i^2 \sigma_i^2 / w^2$.

10.6 Using Taylor's series or otherwise, show that the expectation and variance of $1/w$ are $1/W$ and $V(w)/W^2$ respectively. Using this result, show that the d.f. are as presented in Section 10.5.

10.7 (a) Find the variances of $v_1(\bar{y})$ and $v_2(\bar{y})$, presented in Section 10.6 for the case of $\sigma_\alpha^2 = 0$ and $\sigma_i^2 \equiv \sigma^2$. (b) Show that $v_3(\bar{y})$ in (10.6) is the minimum variance unbiased estimator of $V(\bar{y})$ and find its variance.

10.8 (a) Show that (10.7) provides the logarithm of the likelihood for $(\mu; \sigma_i^2, \sigma_\alpha^2)$. (b) Find the likelihood equations for estimating σ_α^2 and μ when s_i^2 is substituted for σ_i^2. (c) Find the likelihood equations for estimating μ and σ_i^2 when $\sigma_\alpha^2 = 0$ and $\nu_i \equiv \nu$.

CHAPTER 11

Further topics

11.1 Introduction

A number of practical situations are represented through extensions and modifications of the mixed models described in the previous chapters. Some of these models are of the multivariate or nonlinear type and they may also represent observations of replicated experiments or studies. Estimation procedures for these different topics are briefly described in the following sections.

11.2 Prediction of random effects

The mixed model in (6.1) can be expressed as

$$Y = X\beta + U\xi + \xi_0 \, . \tag{11.1}$$

In this expression, ξ and ξ_0 are the vectors of the random effects and residuals which are assumed to be uncorrelated with zero means and dispersion matrices D_1 and D_0 respectively. With these assumptions, Y has mean $X\beta$ and dispersion

$$\Sigma = UD_1U' + D_0 \, . \tag{11.2}$$

Prediction of the individual elements of ξ or a linear combination $p'\xi$ for a specified vector p is of importance for animal and plant improvement programs, scholastic admissions and similar purposes. Cochran (1951) considers the selection of individuals with the largest values of $p'\,\mathrm{E}(\xi|Y)$. Henderson (1975) and Harville (1975) describe the least squares procedure for predicting $q'\beta + p'\xi$ for specified vectors p and q.

To predict $p'\xi$, C.R. Rao (1975) considers

$$p'\widehat{\xi} = C'Y + d \, . \tag{11.3}$$

The vector C and the scalar d are obtained by minimizing the prediction error $E(p'\xi - p'\widehat{\xi})^2$ with the unbiasedness condition $E(p'\widehat{\xi} - p'\xi) = 0$, which implies that $C'X\beta + d = 0$. Through this procedure, the optimum C is found to be

$$C = \Sigma^{-1}UD_1p \, . \tag{11.4}$$

Thus, the best linear unbiased predictor (BLUP) for $p'\xi$ is given by

$$p'\widehat{\xi} = p'D_1U'\Sigma^{-1}(Y - X\beta) \, . \tag{11.5}$$

When ξ and ξ_0 have multinormal distributions, as pointed by C.R. Rao and Kleffé (1988, Section 12.2), the Bayes estimator of ξ is

$$E(\xi|Y) = D_1U'\Sigma^{-1}(Y - X\beta) \, , \tag{11.6}$$

and that of $p'\xi$ is the same as (11.5). This estimator is found by noting that the covariance of ξ and Y is D_1U'.

If Σ is known, $q'\beta + p'\xi$ can be found from replacing β by the least squares estimator $(X'\Sigma^{-1}X)^-X'\Sigma^{-1}Y$ and predicting ξ from (11.6). If Σ is unknown, D_1 and D_0 can be estimated from the procedures in Chapters 6–8.

11.3 Prediction from individual models

Consider the linear models

$$Y_i = X\beta_i + \varepsilon_i \, , \tag{11.7}$$

$i = 1, 2, \ldots, k$, where Y is an m-vector of observations, X is an $(m \times s)$ matrix and β_i is an s-vector of parameters. The m-vector of residuals ε_i is assumed to have mean zero and dispersion matrix $\sigma^2 V$ and is independent of ε_j $(i \neq j)$. These models represent, for instance, responses of k individuals on m occasions.

The least squares estimator of $p'\beta_i$ is $p'\beta_i^{(l)} = p'GX'V^{-1}Y_i$, where $G = (X'V^{-1}X)^-$. Since the dispersion of $\beta_i^{(l)}$ is

$$D_i^{(l)} = E(\beta_i^{(l)} - \beta_i)(\beta_i^{(l)} - \beta_i)' = \sigma^2 G \, , \tag{11.8}$$

the variance of $p'\beta_i^{(l)}$ is $\sigma^2 \, p'Gp$.

If the k units are a random sample from a large population, β_i can be considered to be random with mean γ and dispersion matrix

$\boldsymbol{\Gamma}$. With these assumptions, C.R. Rao (1975) derives the predictor for $\boldsymbol{p}'\boldsymbol{\beta}_i$ from $\boldsymbol{C}'\boldsymbol{Y}_i + d$, as described in the last section. From this procedure, the BLUP or the Bayes estimator for $\boldsymbol{\beta}_i$ is given by

$$\boldsymbol{\beta}_i^{(b)} = \boldsymbol{\beta}_i^{(l)} - \sigma^2 \boldsymbol{G}(\boldsymbol{\Gamma} + \sigma^2 \boldsymbol{G})^{-1}(\boldsymbol{\beta}_i^{(l)} - \boldsymbol{\gamma}) \ . \tag{11.9}$$

The dispersion of $\boldsymbol{\beta}_i^{(b)}$ is

$$
\begin{aligned}
D_b &= E(\boldsymbol{\beta}_i^{(b)} - \boldsymbol{\beta}_i)(\boldsymbol{\beta}_i^{(b)} - \boldsymbol{\beta}_i)' \\
&= \sigma^2 \boldsymbol{G} - \sigma^2 \boldsymbol{G}(\boldsymbol{\Gamma} + \sigma^2 \boldsymbol{G})^{-1}\boldsymbol{G} \ .
\end{aligned}
\tag{11.10}
$$

The prediction error of $\boldsymbol{p}'\boldsymbol{\beta}_i^{(b)}$ is $\boldsymbol{p}'\boldsymbol{D}_b\boldsymbol{p}$, which is clearly smaller than the variance of $\boldsymbol{p}'\boldsymbol{\beta}_i^{(l)}$.

As an illustration, consider the unbalanced one-way model described in Section 2.7. An expression for the BLUP of μ_i can be obtained from (11.9). For a simplified procedure, we may consider the model for the sample means,

$$\overline{y}_i = \mu + \alpha_i + \overline{\varepsilon}_i \ , \tag{11.11}$$

where $\overline{\varepsilon}_i = \sum_{j=1}^{n_i} \varepsilon_{ij}/n_i$ has mean zero and variance σ^2/n_i. Since $V(\overline{y}_i) = \sigma_\alpha^2 + \sigma^2/n_i$ and $Cov(\alpha_i, \overline{y}_i) = \sigma_\alpha^2$, the BLUP of α_i is

$$\hat{\alpha}_i = \frac{\sigma_\alpha^2}{\sigma_\alpha^2 + \sigma^2/n_i}(\overline{y}_i - \mu) \ . \tag{11.12}$$

From the above procedure, the Bayes estimator of μ_i is

$$
\begin{aligned}
\hat{\mu}_i &= \mu + \frac{\sigma_\alpha^2}{\sigma_\alpha^2 + \sigma^2/n_i}(\overline{y}_i - \mu) \\
&= \overline{y}_i - \frac{\sigma^2/n_i}{\sigma_\alpha^2 + \sigma^2/n_i}(\overline{y}_i - \mu) \\
&= \frac{\sigma_\alpha^2}{\sigma_\alpha^2 + \sigma^2/n_i}\overline{y}_i + \frac{\sigma^2/n_i}{\sigma_\alpha^2 + \sigma^2/n_i}\mu \ .
\end{aligned}
\tag{11.13}
$$

As can be seen from the last expression, the weights attached to $\overline{y}_i$ and μ are inversely proportional to σ^2/n_i and σ_α^2. Secondly, $\hat{\mu}_i$ approaches $\overline{y}_i$ as n_i becomes large. The error of this estimator is

$$E(\hat{\mu}_i - \mu_i)^2 = \frac{\sigma_\alpha^2 \sigma^2/n_i}{\sigma_\alpha^2 + \sigma^2/n_i} \ , \tag{11.14}$$

which is smaller than the variance σ^2/n_i of $\overline{y}_i$. For predicting hospital discharges, P.S.R.S. Rao and Shimizu (1991) compared the estimator in (11.13) with other procedures.

For the empirical Bayes estimator, C.R. Rao (1975) substituted unbiased estimators for σ^2, γ and $\boldsymbol{\Gamma}$ in (11.9) and minimized the average dispersion error of the resulting estimator of β_i.

11.4 Growth curves and repeated measures

Growth curves refer to the models employed for studying changes in characteristics over periods of time. Study of the heights and weights of children during their growth years provides an illustration. Examinations of blood pressures, cholesterol levels and similar diagnostic measurements of patients at specified intervals of time provide other illustrations.

The observations for the ith individual at the tth time can be represented by a model of the form

$$y_{it} = \beta_{0i} + \beta_{1i}x_{1t} + \beta_{2i}x_{2t} + \cdots + \varepsilon_{it} . \tag{11.15}$$

In this model, t represents time and $(x_{1t}, x_{2t}, \ldots)$ the covariates which can take the form of $(t, t^2, \ldots,)$. The growth parameters $(\beta_{0i}, \beta_{1i}, \beta_{2i}, \ldots)$ are assumed to be random. For this type of models, the Bayes and empirical Bayes estimators can be found from the procedures in the last section.

To study the long-term effects of air pollution, tobacco smoke and fossil-fuel combustion on pulmonary functions of children, for the ith child, Laird and Ware (1982) consider a model of the form

$$\boldsymbol{Y}_i = \boldsymbol{X}_i\boldsymbol{\beta} + \boldsymbol{U}_i\boldsymbol{\xi}_i + \boldsymbol{\xi}_{oi} . \tag{11.16}$$

The vector of the random effects $\boldsymbol{\xi}_i$ is assumed to have mean zero and dispersion matrix $\boldsymbol{D}$. The vectors of the residuals $\boldsymbol{\xi}_{oi}$ are assumed to have zero means and dispersion matrices $\boldsymbol{D}_i$, and are independent of $\boldsymbol{\xi}_i$. With these assumptions, the dispersion of $\boldsymbol{Y}_i$ is $\boldsymbol{\Sigma}_i = \boldsymbol{U}_i\boldsymbol{D}\boldsymbol{U}_i' + \boldsymbol{D}_i$. As in Section 11.2, the predictor of $\boldsymbol{\xi}_i$ is

$$\widehat{\boldsymbol{\xi}}_i = \boldsymbol{D}\boldsymbol{U}_i'\boldsymbol{\Sigma}_i^{-1}(\boldsymbol{Y}_i - \boldsymbol{X}_i\boldsymbol{\beta}) . \tag{11.17}$$

If $\boldsymbol{D}$ and $\boldsymbol{D}_i$ are known, $\boldsymbol{\beta}$ in this expression is replaced by its least squares estimator $\widehat{\boldsymbol{\beta}} = (\sum_i \boldsymbol{X}_i'\boldsymbol{\Sigma}_i^{-1}\boldsymbol{X}_i)^{-1}(\sum_i \boldsymbol{X}_i'\boldsymbol{\Sigma}_i^{-1}\boldsymbol{Y}_i)$.

Reinsel (1982) and Crowder and Hand (1990), for instance, present illustrations and estimation procedures for the growth curve models. Jennrich and Schluchter (1986) consider structural covariance matrices for models with repeated measures. A review of the literature on this topic is provided by von Rosen (1991).

11.5 Hierarchical models and prediction

The model for the one-way design in Section 2.3 can be considered to be a two-stage model. At the first stage, it is assumed that $E(y_{ij}|\mu_i, \sigma_i^2) = \mu_i$, and at the second stage $E(\mu_i) = \mu$. Similarly, the model in (5.1) for the nested design is a three-stage model. In some situations, assumptions regarding the observations, means and variances are made in stages.

To predict the quality of medical services received by patients, Malec and Sedransk (1985) and Calvin and Sedransk (1991) develop hierarchical models in three stages. General descriptions of the hierarchical models and the related estimation and prediction procedures can be found in C.R. Rao and Kleffé (1988, Ch. 8) and Searle, Casella and McCulloch (1992, Ch. 9). Giesbrecht and Burrows (1983) describe the MINQUE and REML procedures for models with hierarchical structures.

11.6 Multivariate models

The models and the estimation procedures presented in the previous chapters can be extended to the case of two or more correlated characteristics. C.R. Rao and Kleffé (1988) present the MINQUE and other procedures for the multivariate mixed model. Khatri (1979) derived the MIVQUEs for the multivariate analysis of variance model. For illustration, estimation for the one-way classification is presented below.

As in the case of a single character considered in Chapter 2, the observations on $p(> 1)$ characteristics for the k groups can be represented by

$$Y_{ij} = \mu + \alpha_i + \varepsilon_{ij} \, , \tag{11.18}$$

$i = 1, 2, \ldots, k$ and $j = 1, 2, \ldots, m$, where Y, μ, α and ε are p-vectors. The residual vector ε_{ij} is assumed to have mean zero and dispersion matrix Σ_ε. The vector of the random effects α_i is assumed to have mean zero and dispersion matrix $\boldsymbol{\Sigma}_\alpha$, and is independent of $\boldsymbol{\varepsilon}_{ij}$.

The p-vectors of the means of the ith group and the overall mean respectively are $\overline{\boldsymbol{Y}}_i = (1/m) \sum_j \boldsymbol{Y}_{ij}$ and $\overline{\boldsymbol{Y}} = (1/n) \sum_{ij} \boldsymbol{Y}_{ij} = (1/k) \sum_i \overline{\boldsymbol{Y}}_i$, where $n = mk$ is the total sample size. The $p \times p$ matrices of the total, between and within sums of squares and cross-products are respectively given by

$$\boldsymbol{S}_T = \sum_i \sum_j (\boldsymbol{Y}_{ij} - \overline{\boldsymbol{Y}})(\boldsymbol{Y}_{ij} - \overline{\boldsymbol{Y}})' , \tag{11.19}$$

$$\boldsymbol{S}_B = m \sum_i (\overline{\boldsymbol{Y}}_i - \overline{\boldsymbol{Y}})(\overline{\boldsymbol{Y}}_i - \overline{\boldsymbol{Y}})' , \tag{11.20}$$

and

$$\boldsymbol{S}_W = \sum_i \sum_j (\boldsymbol{Y}_{ij} - \overline{\boldsymbol{Y}}_i)(\boldsymbol{Y}_{ij} - \overline{\boldsymbol{Y}}_i)' . \tag{11.21}$$

From these expressions, $\boldsymbol{S}_T = \boldsymbol{S}_B + \boldsymbol{S}_W$. The matrices of the between and within mean squares are $\boldsymbol{M}_B = (k-1)^{-1} \boldsymbol{S}_B$ and $\boldsymbol{M}_W = (n-k)^{-1} \boldsymbol{S}_W$.

With the above assumptions for $\boldsymbol{\alpha}_i$ and $\boldsymbol{\varepsilon}_{ij}$, $E(\boldsymbol{M}_W) = \boldsymbol{\Sigma}_\varepsilon$ and $E(\boldsymbol{M}_B) = m\boldsymbol{\Sigma}_\alpha + \boldsymbol{\Sigma}_\varepsilon$. Thus, unbiased estimators for $\boldsymbol{\Sigma}_\varepsilon$ and $\boldsymbol{\Sigma}_\alpha$ are given by $\hat{\boldsymbol{\Sigma}}_\varepsilon = \boldsymbol{M}_E$ and $\hat{\boldsymbol{\Sigma}}_\alpha = (1/m)(\boldsymbol{M}_B - \boldsymbol{M}_W)$. These are the multivariate analysis of variance (MANOVA) estimators.

For the balanced case, it can be seen that the MIVQUEs of $\boldsymbol{\Sigma}_\varepsilon$ and $\boldsymbol{\Sigma}_\alpha$ are the same as the ANOVA estimators. With the assumption of multivariate normality for $\boldsymbol{\alpha}_i$ and $\boldsymbol{\varepsilon}_{ij}$, the REML estimates are obtained by adjusting these solutions for nonnegative definiteness of $\hat{\boldsymbol{\Sigma}}_\alpha$. Anderson, Anderson and Olkin (1986) present the ML procedure for $\boldsymbol{\Sigma}_\alpha$ and $\boldsymbol{\Sigma}_\varepsilon$. Meyer (1985) presents the REML procedure for the multivariate mixed two-way classification.

11.7 Nonlinear models and variance components

As is the case of a linear model, consider an n-vector of observations $\boldsymbol{Y}$, the $n \times s$ matrix $\boldsymbol{X}$, which is known, and the s-vector of parameters $\boldsymbol{\beta}$. For the logistic regression, it is assumed that $E(y_i | \boldsymbol{X}_i') = e^{\boldsymbol{X}_i'\boldsymbol{\beta}}/(1 + e^{\boldsymbol{X}_i'\boldsymbol{\beta}})$, where y_i is the ith observation and $\boldsymbol{X}_i'$ is the ith row of $\boldsymbol{X}$. In some applications, some of the elements of $\boldsymbol{\beta}$ are considered to be fixed and the remaining random.

Variations of the logistic regression are frequently employed for studying the responses to different dosages of medical treatments. The variance components and fixed effects for these models can be estimated from the procedures described in the previous chapters. Rudemo, Ruppert and Streiberg (1989) considered a logistic model for analyzing data on herbicides. After suitable transformations, the model with both fixed and random effects was found to fit the data satisfactorily.

To analyze data on blood pressures, Solomon and Cox (1992) considered a model of the form

$$y_{ij} = \mu + \alpha_i + \varepsilon_{ij} + a_{20}\alpha_i^2 + a_{11}\alpha_i\varepsilon_{ij} + a_{02}\varepsilon_{ij}^2 \ . \qquad (11.22)$$

In this model, μ is the overall mean, (a_{20}, a_{11}, a_{02}) are fixed parameters, α_i is the random effect and ε_{ij} is the residual. These authors also developed an approximate likelihood function for the nonlinear variance components model, followed by a test for examining the difference between two treatments.

Estimation procedures for the nonlinear mixed models were suggested, for instance, by Davidian and Giltinan (1995). Lindstrom and Bates (1990) and Vonesh and Carter (1992) consider these models for repeated measures.

11.8 Generalized linear models with random effects

For the logistic regression of a proportion p, the logit $\ln[p/(1-p)]$ is represented by a linear model. Drum and McCullagh (1993) present the REML estimation for the logistic model. McCullagh and Nelder (1989) describe similar transformations for models with quantitative or qualitative variables and present the estimation procedures. Extensions and modifications of these procedures have been considered for models that include both fixed and random effects. Reinsel (1984), for instance, presents the estimation procedure for a multivariate generalized linear model with random effects.

Exercises

11.1 Show that the BLUP for $p'\xi$ takes the form of (11.5) and find its error, $E(p'\widehat{\xi} - p'\xi)^2$, when the least squares estimator is substituted for β. Assume that Σ is known.

11.2 For the model in (11.7), show that the BLUP and Bayes estimators take the form of (11.9).

11.3 Without the simplification in (11.11), show that the estimators (11.12) and (11.13) can be obtained directly from the procedure in Section 11.2 or 11.3.

11.4 Consider the BLUP estimator in (2.28) and derive the empirical Bayes estimator for μ_i through C.R. Rao's (1975) procedure described in Section 11.3.

11.5 Find the dispersion error of the predictor in (11.17) when the least squares estimator is substituted for β.

11.6 For the balanced one-way model described in Section 11.6, show that the MIVQUEs and the solutions to the REML equations are the same as the ANOVA estimators.

11.7 Describe the procedures for obtaining the MIVQUEs of Σ_ε and Σ_α for the unbalanced case of the model in (11.18).

Solutions to exercises

Chapter 2

2.1 (a) $\hat{\sigma}^2 = 750.15$, S.E. $(\hat{\sigma}^2) = 187.54$; $\hat{\sigma}_\alpha^2 = 567.45$; S.E.$(\hat{\sigma}_\alpha^2) = 449.8$; $\hat{\mu} = 308.04$; S.E.$(\hat{\mu}) = 14.56$.
(b) Confidence limits for σ^2, $\sigma_{alpha}^2/\sigma^2$ and μ respectively are $(479.02, 1340.35)$, $(0.47, 0.84)$ and $(245.39, 370.69)$.

2.2 (a) $\hat{\sigma}_{\alpha d}^2 = 2.07$, $\hat{\sigma}_d^2 = 1.34$ and $\hat{\mu}_d = 1.99$.
(b) S.E.s: 1.60, 0.42 and 0.87.
(c) $t_2 = 2.29$ significant at the 0.08 level.

2.3 (a) $\hat{\sigma}^2 = 1.24$
(b) $\hat{\sigma}_i^2 = 0.883$, 2.5 and 0.916.
$\hat{\sigma}_\alpha^2$ is negative in both cases.

2.4 $\hat{\sigma}_\alpha^2 = 1.18$ and S.E.$(\hat{\sigma}_\alpha^2) = 1.45$.

2.5 $\hat{\beta} = 0.39$. To test $\beta = 0$, $F(1, 11) = 7.34$ which is significant at the 0.05 level. Without the covariate, $\bar{y}_2 - \bar{y}_1 = 20$, $\bar{y}_3 - \bar{y}_2 = 3$, and $\bar{y}_3 - \bar{y}_1 = 23$ with a S.E. of 2.38 for each difference. With the covariate adjustment, $\bar{y}_3 - \bar{y}_2 = -1.67$, $\bar{y}_3 - \bar{y}_1 = 10.17$ and $\bar{y}_2 - \bar{y}_1 = 11.83$ with S.E.s 2.58, 5.11 and 3.58 respectively.

2.6 $\hat{\sigma}_\alpha^2 = 20.9$.

2.11 $\hat{\sigma}_\alpha^2 = (M_B - M_W)/m$ is distributed approximately as a χ^2-distribution with ν d.f., where $\nu = 2\sigma_\alpha^4/V(\hat{\sigma}_\alpha^2)$.

2.12 $(F - F_u)/[F + (m-1)F_u] \le \rho \le (F - F_l)/[F + (m-1)F_l]$.

2.13 (a) $\hat{T} = [M_B + (m-1)M_W]/m$.
(b) $V(\hat{T}) = (2/m^2)[(m\sigma_\alpha^2 + \sigma^2)/(k-1) + (m-1)^2\sigma^4]/(n-k)$.

(c) $v(\hat{T}) = (2/m^2)[M_B^2/(k+1) + (m-1)^2 M_W^2/(n-k+2)]$.

(d) $\hat{T}$ is approximately distributed as χ^2 with $\nu = 2T^2/V(\hat{T})$ d.f.

Chapter 3

3.1 $0 < \sigma_\alpha^2/\sigma^2 < 4.04$.

3.2 $\hat{\sigma}_\alpha^2 = 15.97$ and S.E.$(\hat{\sigma}_\alpha^2) = 11.18$.

3.3 $0.06 < \sigma_\alpha^2/\sigma^2 < 17.53$.

3.4 At the 0.05 level, power when $\sigma_\alpha^2/\sigma^2 = 1$, 2 and 10 is given by $F(4, 23)$ exceeding 2.8/7, 2.8/13 and 2.8/61 respectively.

3.5 (a) $\hat{\sigma}_\alpha^2 = 817.36$ and $\hat{\sigma}_\beta^2 = 85.24$.
(b) S.E.$(\hat{\sigma}_\alpha^2) = 529.59$ and S.E.$(\hat{\sigma}_\beta^2) = 71.19$.

3.6 (a) $\hat{\sigma}_\alpha^2 = 1812.78$ and S.E.$(\hat{\sigma}_\alpha^2) = 908.8$.
(b) $\hat{\sigma}_\beta^2 = 67{,}745$ and S.E.$(\hat{\sigma}_\beta^2) = 27{,}660$.

3.7 $F(6,12) = 119.1/109.38 = 1.09$ is not significant at the 0.05 level.

3.8 $F(1,5) = 128.24/45.81 = 2.8$ is not significant at the 0.05 level.

3.9 $\hat{\sigma}_\gamma^2 = 6.36$ and S.E.$(\hat{\sigma}_\gamma^2) = 36.25$.

3.12 $F(6, 20) = 3.48$ is significant at the 0.05 level.

Chapter 4

4.1 (a) $F(3, 6) = 4.49$
(b) $\hat{\sigma}_t^2 = 19.94$ and S.E.$(\hat{\sigma}_t^2) = 16.48$.

4.2 Percentage decline: For the systolic pressures, $F(3, 6) = 54.3$ for the treatments which is highly significant. For the diastolic pressures, $F(3, 6) = 3.22$ for the treatments, which is not significant at the 0.05 level.

4.3 1.75 and 1.10.

4.4 -2 and 9.47.

4.5 1.70 and 1.07.

4.6 Adjusted totals for A, B, C and D respectively are $-32/3$, $-28/3$, $48/3$ and 4. Adjusted treatment effects are obtained by multiplying these figures by $3/8$. S_B(unadj.) $= 38.67$, S_T(unadj.) $= 204$, S_T(adj) $= 177.33$, S_B(adj) $= 12$ and $S_E = 30.67$. $F(3, 5) = M_T/M_E = 9.64$. S.E. of the difference of the two treatment effects is 2.15.

4.7 (a) For the blocks (drivers), $F(3, 9) = 1.58/7.58$ which is insignificant. (b) $\hat{\sigma}_\alpha^2 = 0$.

4.8 MSEs for treatments, replications, rows in replications, columns in replications and residual are 142.46, 15.13, 7.96, 8.96 and 6.93 with 3, 1, 6, 6 and 15 d.f. respectively. For the treatments $F(3, 15) = 142.46/6.93 = 20.57$ is highly significant.

4.9 $V_1 = V(\sum_j c_j \hat{\tau}_j) = \sigma^2 \sum_j c_j^2/rE$ and $V_2 = V(\sum_j c_j \tilde{\tau}) = k(k\sigma_b^2 + \sigma^2)\sum_j c_j^2/(r - \lambda)$. The combined estimator is $w \sum_j c_j \hat{\tau}_j + (1 - w)\sum_j c_j \tilde{\tau}$. The value of w and the expression for the variance for this estimator take the same forms as presented in Section 4.5.2.

Chapter 5

5.1 MSEs for regions (A), states (B) in regions and the residual are 733.63, 368.75 and 156.72 with 2, 3 and 18 d.f. respectively.
(1) When both the regions and states are fixed, for the regions $F(2, 18) = M_A/M_E = 4.68$ is significant at the 0.05 level and for the states $F(3, 18) = M_B/M_E = 2.35$ is not significant at the 0.05 level.
(2) If regions are fixed and states are random, for the regions, $F(2, 3) = M_A/M_B = 1.99$ is not significant at the 0.05 level. For the states, $\hat{\sigma}_\beta^2 = 53.01$ and S.E.$(\hat{\sigma}_\beta^2) = 59.61$.
(3) If both the regions and states are random, for the regions, $\hat{\sigma}_\alpha^2 = 45.61$ and S.E.$(\hat{\sigma}_\alpha^2) = 71.1$. For the states, $\hat{\sigma}_\beta^2 = 53.01$ and S.E.$(\hat{\sigma}_\beta^2) = 59.61$.

5.2 In this case, $M_A = 255.79$ and $M_E = 31.73$, $\hat{\sigma}_\alpha^2 = 28.01$, S.E.$(\hat{\sigma}_\alpha^2) = 22.64$ and $F(2, 21) = 8.06$ which is significant at the 0.01 level.

5.3 Denoting the fabrics, tests and repetitions by A, B and C, $a =$

4, $b = 3$ and $c = 4$. The estimating equations are $bc\hat{\sigma}_\alpha^2 + c\hat{\sigma}_\beta^2 + \hat{\sigma}^2$ $= 17.54$, $c\hat{\sigma}_\beta^2 + \hat{\sigma}^2 = 30.53$ and $\hat{\sigma}^2 = 7.54$. Thus, for the tests, $\hat{\sigma}_\beta^2$ $= 5.52$, S.E.$(\hat{\sigma}_\beta^2) = 3.44$ and $F(8, 36) = 4.05$ which is significant at the 0.05 level. The estimate for σ_α^2 is negative.

5.4 Denoting the stalks, branches, nodes and cladophils by A, B, C and D, their numbers are $a = 5$, $b = 2$, $c = 5$ and $d = 3$. The estimating equations are $abc\hat{\sigma}_\alpha^2 + bc\hat{\sigma}_\beta^2 + c\hat{\sigma}_\gamma^2 + \hat{\sigma}^2 = 101.86$, $bc\hat{\sigma}_\beta^2 + c\hat{\sigma}_\gamma^2 + \hat{\sigma}^2 = 21.11$, $c\hat{\sigma}_\alpha^2 + \hat{\sigma}^2 = 5.23$ and $\hat{\sigma}^2 = 3.35$.
(1) For the stalks, $\hat{\sigma}_\alpha^2 = 1.62$ and $F(4, 5) = 101.86/21.11 = 4.83$ is significant at the 0.06 level.
(2) For the branches, $\hat{\sigma}_\beta^2 = 1.59$ and $F(5, 40) = 4.04$ is significant at the 0.01 level.
(3) For the nodes, $\hat{\sigma}_\gamma^2 = 0.38$ and $F(40, 100) = 5.23/3.35 = 1.56$ is significant at the 0.05 level.

5.6 Denoting the operators, samples, runs and tests by A, B, C and D, their numbers are $a = 2$, $b = 2$, $c = 3$ and $d = 2$. The estimating equations are $abc\hat{\sigma}_\alpha^2 + bc\hat{\sigma}_\beta^2 + c\hat{\sigma}_\gamma^2 + \hat{\sigma}^2 = 4510.04$, $bc\hat{\sigma}_\beta^2 + c\hat{\sigma}_\gamma^2 + \hat{\sigma}^2$ $= 69.38$, $c\hat{\sigma}_\gamma^2 + \hat{\sigma}^2 = 27.21$, and $\hat{\sigma}^2 = 16.3$. Hence, $\sigma_\alpha^2 = (4510.04 - 69.38)/12 = 370.55$, and $\sigma_\beta^2 = (69.38 - 27.21)/6 = 7.03$

Chapter 9

9.1 (1) (0.75, 17.5). (2) (0.87, 10.16)

9.2 $\hat{T} = 2.1$. For the MLS approximation, $0.37 \leq T \leq 31.78$.

9.3 $\hat{\sigma}_\alpha^2 = 0.91$ and $\hat{T} = 1.95$. For the χ^2-approximation, $0.92 \leq T \leq 6.5$, and for the normal approximation, $0.09 \leq T \leq 3.81$.

9.4 Limits for σ_α^2 from (9.10) are too wide and the limits for the normal approximation are too narrow.
(a) (0.32, 46.8) and
(c) (0.16, 51.2).

9.6 $-0.14 \leq \rho \leq 0.98$.

Chapter 10

10.1 $\hat{\sigma}_\alpha^2$ is negative.

10.2 $\overline{y}_W = 1.91$, S.E.$(\overline{y}_W) = 0.64$, and confidence limits for the mean are $(0.65, 3.16)$. $F(4, 37) = 1.04$ is not significant. $\hat{\sigma}_\alpha^2 = 0.18$, $\overline{y}_w = 1.75$ and S.E.$(\overline{y}_w) = 0.59$.

10.3 At $x_i = 10$, $\hat{y}_i = 17.48$, S.E.$(\hat{y}_i) = 1.51$, and the confidence limits are $(14.52, 20.44)$. At $x_i = 30$, $\hat{y}_i = 40.28$, S.E.$(\hat{y}_i) = 4.24$ and the confidence limits are $(31.97, 48.59)$.

10.4 (a) $\overline{y} = 0.55$, $s^2 = 0.0076$, $\hat{\sigma}_\alpha^2 = 0.10$, $\overline{y}_w = 0.48$, S.E.$(\overline{y}_w) = 0.13$, and 95 percent confidence limits for the mean are $(0.22, 0.74)$.
(b) $F = 14.2$ with 2 and 12 d.f., which is highly significant.

Bibliography

Ahrens, H., Kleffé, J. and Tenzler, R. (1981). Mean Square Error Comparison of MINQUE, ANOVA and two Alternative Estimators under the Unbalanced One-way Random Effects Model. *Biometrical Journal* **23**, 323–342.

Anderson, B.M., Anderson, T.W. and Olkin, L.(1986). Maximum Likelihood Estimators and Likelihood Criteria in Multivariate Components of Variance. *Annals of Statistics* **14**, 405–417.

Anderson, R.L. and Bancroft, T.A. (1952). *Statistical Theory in Research*. McGraw Hill Book Co., New York.

Anderson, R.L. and Crump, P.P. (1967). Comparison of Designs and Estimation Procedures in a Two-stage Nested Process. *Technometrics* **9**, 499–516.

Bainbridge, T.R. (1965). Staggered Nested Designs for Estimating Variance Components. *Industrial Quality Control* **22**, 12–20.

Bancroft, T.A. (1944). On Biases in Estimation due to the use of Preliminary Tests of Significance. *Annals of Mathematical Statistics* **15**, 190–204.

Begg, C.B. and Pilote, L. (1991). A Model for Incorporating Controls into a Meta-Analysis. *Biometrics* **47**, 899–906.

Blischke, W.R. (1966). Variances of Estimates of Variance Components in Three-way Classification. *Biometrics* **22**, 553–565.

Boardman, T.J. (1974). Confidence Intervals for Variance Components - A Comparative Monte Carlo Study. *Biometrics* **30**, 251–262.

Bremer, R. (1990). Numerical study of the small sample variances of the MINQUE estimators of variance components in the two-way factorial design. *Communications in Statistics* **A19**, 1261–1279.

Broemeling, L.D. (1969). Confidence Intervals for Variance Ratios of Random Models. *Journal of the American Statistical Association* **64**, 660–664.

Bross, I.D.J. (1950). Fiducial Intervals for Variance Components. *Biometrics* **6**, 136–140.

Brown, K.G. (1977). On Estimation of Diagonal Covariance Matrices by MINQUE. *Communications in Statistics* **A 6**(5), 471–484.

Brown, K.G. and Burgess, M.A. (1984). On Maximum Likelihood and Restricted Maximum Likelihood Approaches to Estimation of Variance Components. *Journal of Statistical Computation and Simulation* **19**, 59–77.

Brown, S.S., Healy, M.J.R. and Kearns, M. (1981). Report on the Interlaboratory Trial of the Reference Method for the Determination of Total Calcium in Serum. *Journal of Clinical Chemistry and Clinical Biochemistry* **19**, 395–426.

Brownlee (1965). *Statistical Theory and Methodology in Science and Engineering*, Second Edition. John Wiley and Sons, New York.

Bulmer, M.G. (1957). Approximate Confidence Limits for Components of Variance. *Biometrika* **44**, 159–167.

Burdick, R.K. and Eickman, J. (1986). Confidence Intervals on the Among group Variance Component in the Unbalanced One-fold nested Design. *Journal of Statistical Computation and Simulation* **26**, 205–219.

Burdick, R.K. and Graybill, F.A.(1992). *Confidence Intervals on Variance Components*. Marcel Dekker, New York.

Burdick, R.K. and Sielken, R. L. Jr. (1978). Exact Confidence Intervals for Linear Combinations of Variance Components in Nested Classifications. *Journal of the American Statistical Association* **73**, 632–635.

Burdick, R.K. Maqsood, F. and Graybill, F.A.(1986). Confidence Intervals on the Intraclass Correlation in the Unbalanced One-way Classification. *Communications in Statistics* **A 15**, 3353–3378.

Callanan, T. P. and Harville, D. A. (1991). Some new Algorithms for Computing Restricted Maximum Likelihood Estimates of Variance Components. *Journal of Statistical Computation and Simulation* **38**, 239–259.

Calvin, J.A. and Sedransk, J. (1991). Bayesian and Frequentist Predictive Inference for the Patterns of Health Care Studies. *Journal of the American Statistical Association* **86**(413), 36–48.

Cameron, J.M. (1951). Use of Components of Variance in Preparing Schedules for the Sampling of Baled Wool. *Biometrics* **7**, 83–96.

Canner, P.L., Borhani, N.M., Oberman, A., Cutler, J., Prineas, R.J., Langford, H. and Hopper, F. (1991). The Hypertension Prevention Trial: Assessment of the Quality of Blood Pressure Measurements. *American Journal of Epidemiology* **134**(4), 379–391.

Chamberlain, A.C. and Turner, F.M. (1952). Errors and Variations in White-cell Counts. *Biometrics* **8**, 55–65.

Chaubey, Y.P. (1980). Application of the Method of MINQUE for Estimation in Regression with Intraclass Covariance Matrix. *Sankhya* **B 42**(1&2), 28–32.

Chaubey, Y.P. (1983). A Non-negative Estimator of Variance Component Closest to MINQUE. *Sankhya* **A 45**, 201–211.

Chaubey, Y.P. (1984). On the Comparison of some Non-negative Estimators of Variance Components of two Models. *Communications in Statistics, Simulation and Computation* **13**, 619–633.

Chaubey, Y.P and Rao, P.S.R.S. (1976). Efficiency of Five Estimators for the Parameters of Two Linear Models with Unequal Variances. *Sankhya* **B 38**, 364–370.

Chew, V. (1970). Covariance Matrix Estimation in Linear Models. *Journal of the American Statistical Association* **65**, 173–181.

Cochran, W.G. (1934). The Distribution of Quadratic forms in a Normal System. *Proceedings of the Cambridge Philosophical Society*, Supplement **4**, 102–118.

Cochran, W.G. (1937). Problems Arising in the Analysis of a Series of Similar Experiments. *Journal of the Royal Statistical Society*, Supplement 4, 102–118.

Cochran, W.G. (1938). Long term Agricultural Experiments. *Journal of the Royal Statistical Society*, Supplement **6**, 104–148.

Cochran, W.G. (1939). The use of Analysis of Variance in Enumeration by Sampling. *Journal of the American Statistical Association* **34**, 492–510.

Cochran, W.G. (1946). Analysis of Covariance. Paper Presented at the Institute of Mathematical Statistics Meetings; Princeton University.

Cochran, W.G. (1951). Improvement by means of Selection. *Proceedings of the Second Berkeley Symposium in Mathematical Statistics and Probability*, 449–470. University of California Press.

Cochran, W.G. (1951). Testing a Linear Relation Among Variances. *Biometrics* **7**, 17–32.

Cochran, W.G. (1954). The Combination of Estimates from Different Experiments. *Biometrics* **10**, 109–129.

Cochran, W.G. and Carroll, S.P. (1953). A Sampling Investigation of the Efficiency of Weighting Inversely as the Estimated Variance. *Biometrics* **9**, 447–459.

Cochran, W.G. and Cox, G.M. (1957). *Experimental Designs*. John Wiley and Sons, New York.

Conerly, M.D. and Webster, J.T. (1987). MINQE for the one-way Classification. *Technometrics* **29**(2), 229–236.

Corbeil, R.R. and Searle, S.R. (1976). A Comparison of Variance

Component Estimators. *Biometrics* **32**, 779–791.

Cox, D. R. (1958). *Planning of Experiments*. John Wiley and Sons, New York.

Crowder, M. (1992). Interlaboratory Comparisons: Round Robins with Random Effects. *Appplied Statistics* **41**(2), 409–425.

Crowder, M.J. and Hand, D.J. (1990). *Analysis of Repeated Measures*. Monographs on Statistics and Applied Probability. Chapman and Hall, London and New York.

Crump, S.L. (1946). The Estimation of Variance Components in Analysis of Variance. *Biometrics Bulletin* **2**, 7–11.

Crump, S.L. (1951). The Present Status of Variance Component Analysis. *Biometrics* **7**, 1–16.

Cuttler, J.A., Follmann, D., Elliott, P. and Suh, Il. (1991). An Overview of Randomized Trials of Sodium Reduction and Blood Pressure. *Hypertension*, Supplement **17**(1), 27–34.

Daniels, H.E. (1938). Some Problems of Statistical Interest in Wool Research. *Journal of the Royal Statistical Society*, Supplement **5**, 89–128.

Daniels, H.E. (1939). The Estimation of Components of Variance. *Journal of the Royal Statistical Society*, Supplement **6**, 186–197.

Darwin, C. (1859). *The Origin of Species*. Mentor Publication (1964), The New American Library, New York.

Davenport, J.M. (1975). Two Methods of Estimating Degrees of freedom of an Approximate F. *Biometrika* **62**, 682–684.

Davenport, J.M. and Webster, J.T. (1973). A Comparison of some Approximate F-tests. *Technometrics* **15**, 779–789.

Davidian , M. and Giltinan, D. M. (1995). *Nonlinear Models for Repeated Measurement Data*. Monographs on Statistics and Applied Probability. Chapman and Hall, London and New York.

Davies, O.L. (Ed). (1967). *The Design and Analysis of Industrial*

Experiments, 4th Ed. Hafner Publications, New York.

Dempster, A.P., Rubin, D.B. and Tsutakawa, R.K.(1981). Estimation in Covariance Components Models. *Journal of the American Statistical Association* **76**, 341–353.

Der Simonian, R. and Laird, N. (1986). Meta-Analysis in Clinical Trials. *Controlled Clinical Trials* **7**,177–188.

Donner, A. (1986). A Review of Inference Procedures for the Intraclass Correlation Coefficient in the One-way Random Effects Model. *International Statistical Review* **54**, 67–82.

Donner, A. and Wells, G. (1986). A Comparison of Confidence Interval Methods for the Intraclass Correlation Coefficient. *Biometrics* **42**, 401–412.

Drum, M.L. and McCullagh, P. (1993). REML Estimation with Exact Covariance in the Logistic Model. *Biometrics* **49**(3), 677–689.

Dunn, O.J. and Clark, V.A. (1974). *Applied Statistics: Analysis of Variance and Regression*. John Wiley and Sons, New York.

Eisenhart, C. (1947). The Assumptions Underlying the Analysis of Variance. *Biometrics* **3**, 1–21.

Eliasziw, M. and Donner, A. (1990). Comparison of recent Estimators of Interclass correlation from Familial data. *Biometrics* **46**, 391–398.

Falconer, D. S. (1989). *Introduction to Quantitative Genetics*, Third Edition. Longman Scientific and Technical, London.

Fisher, R.A. (1918). The Correlation Between Relatives on the Supposition of Mendelian Inheritance. *Transactions of the Royal Society*, Edinburgh, **52**, 399–433.

Fisher, R.A. (1925). *Statistical Methods for Research Workers*, First Edition, Oliver and Boyd, London.

Fisher, R.A. and Yates, F. (1967). *Statistical Tables for Biological, Agricultural and Medical Research*, Sixth Edition. Oliver and Boyd,

Edinburgh and London.

Fry, J. D. (1992). The Mixed-Model Analysis of Variance Applied to Quantitative Genetics: Biological Meaning of the Parameters. *Evolution* **46**(2), 540–550.

Giesbrecht, F.G. (1983). An Efficient Procedure for Computing MINQUE of Variance Components and Generalized Least Squares Estimates of Fixed Effects. *Communications in Statistics* **A 12**, 2169–2177.

Giesbrecht, F.G. and Burrows, P.M. (1978). Estimating Variance Components in Hierarchical Structures using MINQUE and Restricted Maximum Likelihood. *Communications in Statistics* **A 7**, 891–904.

Goldsmith, C.H. and Gaylor, D.W. (1970). Three Stage Nested Design for Estimating Variance Components. *Technometrics* **12**, 487–498.

Goodnight, J.H. and Hemmerle, W.J. (1979). A Simplified Algorithm for the W-transformation in Variance Component Estimation. *Technometrics* **21**, 265–268.

Graybill, F.A. (1961). *An Introduction to Linear Statistical Models*. Vol. 1. Mc Graw Hill Book Co., New York.

Graybill, F.A. and Wang, C.M. (1980). Confidence Intervals on Nonnegative Linear Combinations of Variances. *Journal of the American Statistical Association* **75**, 869–873.

Grubbs, F.E. (1948). On Estimating Precision of Measuring Instruments and Product Variability. *Journal of the American Statistical Association* **43**, 243–264. Corrigenda, **43**, 564.

Haines,T., Komov,V. and Jagoe, C.H. (1992). Lake Acidity and Mercury Content of Fish in Darwin National Reserve, Russia. *Environmental Pollution* **78**, 107–112.

Halvorsen, K.T. (1984). Estimating Population Parameters Using Information from Several Independent Studies. Unpublished Doctoral Dissertation. Harvard School of Public Health.

Han, C.P. (1978). Nonnegative and Preliminary Test Estimators of Variance Components. *Journal of the American Statistical Association* **73**, 855–858.

Hartley, H.O. and Rao, J.N.K. (1967). Maximum Likelihood Estimation for the Mixed Analysis of Variance Model. *Biometrika* **54**, 93–108.

Hartley, H.O. and Rao, J.N.K. and La Motte, L.R. (1978). A Simple 'Synthesis' Based Method of Variance Components Estimation. *Biometrics* **34**, 233–242.

Hartung, J. (1981). Nonnegative Minimum Biased Invariant Estimation in Variance Component Models. *Annals of Statistics* **9**, 278–292.

Harville, D.A. (1976). Extension of the Gauss–Markov Theorem to Include the Estimation of the Random Effects. *Annals of Statistics*, **4**(2), 384–395.

Harville, D.A. (1977). Maximum Likelihood Approaches to Variance Components Estimation and to Related Problems. *Journal of the American Statistical Association* **72**, 320–340.

Harville, D.A. and Fenech, A.P. (1985). Confidence Intervals for a Variance Ratio, or for Heritability in an Unbalanced Mixed Linear Model. *Biometrics* **41**, 137–152.

Healy, M.J.R. and Kearns, M. (1981). Report on the Interlaboratory Trial of the Reference Method for the Determination of Total Calcium in Serum. *Journal of Clinical Chemistry and Clinical Biochemistry* **19**, 395–426.

Heckler, C. E. (1989). Estimation of the Mean and the Variance Components in the Three-Fold Nested Random Effects Model. Unpublished Doctoral Dissertation, University of Rochester.

Heckler, C.E. and Rao, P.S.R.S. (1985). Efficient Estimation of Variance Components using Staggered Nested Designs. *Fall Technical Conference*, American Society for Quality Control, Corning, New York.

Heine, B. (1993). Nonnegative Estimation of Variance Components in an Unbalanced One Way Random Effects Model. *Communications in Statistics* **A 22**(8), 2351–2372.

Hemmerle, W.J. and Hartley, H.O. (1973). Computing Maximum Likelihood for the Mixed A. O. V. Estimation Using the W Transformation. *Technometrics* **15**, 819–831.

Henderson, C.R. (1953). Estimation of Variance and Covariance Components. *Biometrics* **9**, 226–252.

Henderson, C. R. (1975). Best Linear Unbiased Estimation and Prediction under a Selection Model. *Biometrics* **31**, 423–447.

Herbach, L.H. (1959). Properties of Model II Type Analysis of Variance Tests, A : Optimum Nature of the F-test for Model II in the Balanced Case. *Annals of Mathematical Statistics* **30**, 939–959.

Hess, J.L.(1979). Sensitivity of the MINQUE with Respect to *a priori* Weights. *Biometrics* **35**, 645–649.

Hocking, R.R. (1973). A Discussion of the Two-way Mixed Model. *American Statistician* **27**,148–152.

Horn, S.D. and Horn, R.A. and Duncan, D.B.(1975). Estimating Heteroscedastic Variances in Linear Models. *Journal of the American Statistical Association* **70**, 380–385.

Howe, R.B. and Myers, R.H. (1970). An Alternative to Satterthwaite's Test Involving Positive Linear Combinations of Variance Components. *Journal of the American Statistical Association* **65**, 404–412.

Huitson, A. (1955). A Method of Assigning Confidence Limits to Linear Combinations of Variances. *Biometrika* **42**, 471–479.

Jackson, J.E. and Lawton, W.H. (1969). Comparison of ANOVA and Harmonic Components of Variance. *Technometrics* **1**(1), 75–90.

Jacquez, J.A., Mather, F.J.and Crawford, C.R. (1968). Linear Regression with Nonconstant Error Variances: Sampling Experiments with Least Squares, Weighted Least Squares and Maximum Likelihood eEtimators. *Biometrics* **24**, 607–627.

Jennrich, R.I. and Sampson, P. F. (1976). Newton–Raphson and related Algorithms for Maximum Likelihood Variance Component Estimation. *Technometrics* **18**(1), 11–17.

Jennrich, R.I. and Schluchter, M.D. (1986). Unbalanced Repeated Measures Models with Structural Covariance Matrices. *Biometrics* **42**, 805–820.

John, W.M. (1971). *Statistical Design and Analysis of Experiments*. MacMillan Co., New York.

Johnson, N.L. (1948). Alternative Systems in the Analysis of Variance. *Biometrika* **35**, 80–87.

Kackar, R.N. and Harville, D.A. (1984). Approximation for Standard Errors of Estimators of Fixed and Random Effects in Mixed Linear Models. *Journal of the American Statistical Association* **79**, 853–862.

Kaplan, J.S. (1983). A Method for Calculating MINQUE Estimators of Variance Components. *Journal of the American Statistical Association* **78**, 476–477.

Kelly, R.J. and Mathew, T. (1993). Improved Estimators of Variance Components with Smaller Probability of Negativity. *Journal of the Royal Statistical Society* **B** **55**(4), 897–911.

Kempthorne, O. (1952). *The Design and Analysis of Experiments*. John Wiley and Sons, New York.

Kempthorne, O. (1963). *An Introduction to Genetic Statistics*. John Wiley and Sons, New York.

Khatri, C.G. (1979). Minimum Variance Quadratic Unbiased Estimate of a Linear Function of Variances and Covariances under MANOVA Model. *Journal of Statistical Planning and Inference* **3**, 299–303.

Khuri, A.I. (1981). Simultaneous Confidence Intervals for Functions of Variance Components in Random Models. *Journal of the American Statistical Association* **76**, 878–885.

Khuri, A.I. and Sahai, H. (1985). Variance Components Analysis: A Selective Literature Survey. *International Statistical Review* **53**, 279–300.

Kleffé (1993). Parent-Offspring and Sibling Correlation Estimation based on MINQUE Theory. *Biometrika* **80**(2), 393–404.

Kleffé, J. and Rao, J.N.K. (1986). The Existence of Asymptotically Unbiased Nonnegative Quadratic Estimates of Variance Components in ANOVA Models. *Journal of the American Statistical Association* **81**, 692–698.

Kleffeé, J. and Siefert, B. (1990). Matrix free Computation of C. R. Rao's MINQUE for Unbalanced Nested Classification Models. *Computational Statistics and Data Analysis* **2**, 215–228.

Kleffé, J. Prasad, N.G.N. and Rao, J.N.K. (1991). 'Optimal' Estimation of Correlated Response Variance under Additive Model. *Journal of the American Statistical Association* **86**, 144–153.

Klotz, J.H., Milton, R.C. and Zacks, S. (1969). Mean Square Efficiency of Estimators of Variance Components. *Journal of the American Statistical Association* **64**, 1383–1402.

Koch, C.G., Amara, I.A., Brown Jr. B.A., Colton,T. and Gillings, D.B. (1989). A Two-period Cross Over Design for the Comparison of Two Active Treatments and a Placebo. *Statistics in Medicine* **8**(4), 487–504.

Kronberg, D., Hansen, B. and Aaby, P. (1992). Analysis of the Incubation Period for Measles in the Epidemic in Greenland in 1951 using a Variance Components Model. *Statistics in Medicine* **2**, 579–590.

Kussmaul, K. and Anderson, R.L.(1967). Estimation of Variance Components in Two-stage Nested Designs with Composited Samples. *Technometrics* **9**, 373–389.

Laird, N.M. (1982). Computation of Variance Components using the EM Algorithm. *Journal of Statistical Computation and Simulation* **14**, 295–303.

Laird, N.M. and Ware, J.H. (1982). Random-effects Models for Longitudinal Data. *Biometrics* **38**, 963–974.

LaMotte, L.R. (1973). On Non-negative Quadratic Estimation of Variance Components. *Journal of the American Statistical Association* **68**, 728–730.

LaMotte, L.R. (1973). Quadratic Estimation of Variance Components. *Biometrics* **29**, 311–330.

Lee, K.R. and Kapadia, C.H. (1984). Variance Components Estimators for the Balanced Two-way Mixed Model. *Biometrics* **40**, 507–512.

Leone, F.C., Nelson, L.S., Johnson, N.L. and Eisenstat., S. (1968). Sampling Distributions of Variance Components, II. Empirical Studies of Unbalanced Nested Designs. *Technometrics* **10**, 719–738.

Lindstrom, M.J. and Bates, D.M. (1990). Nonlinear Mixed Effects Models for Repeated Measures Data. *Biometrics* **46**(3), 673–687.

Loh, W-Y. (1986). Improved Estimators for the Ratios of Variance Components. *Journal of the American Statistical Association* **81**, 699–702.

Lou, L-M. and Senturia, J. (1977). Computation of Minque Variance Components Estimates. *Journal of the American Statistical Association* 72, 867–868.

Lu, T-F,C., Graybill, F. A. and Burdick, R. K. (1989). Confidence Intervals on the Ratio of Expected Mean Squares $(q_1 + dq_2)/q_3$. *Biometrics* **43**, 535–543.

Maddala, G.S. and Mount, T.D. (1973). A Comparative Study of Alternative Estimators of Variance Components Models used in Econometric Applications. *Journal of the American Statistical Association* **68**, 324–328.

Mahamunulu, D.M. (1963). Sampling Variances of the Estimates of

Variance Components in the Unbalanced Three-way Classification. *Annals of Mathematical Statistics* **34**, 521–527.

Malec, D. and Sedransk, J. (1985). Bayesian Methodology for Predictive Inference for Finite Population Parameters in Multistage Cluster Sampling. *Journal of the American Statistical Association* **80**, 897–902.

Marshall, R.J. and Mardia, K.V. (1985). Minimum Norm Quadratic Estimation of Components of Spatial Covariance. *Mathematical Geology*, **17**(5), 517–525; 1986, **18**(2), 267.

Mason, R.L, Gunst, R.F. and Hess, J.L. (1989). *Statistical Design and Analysis of Experiments with Applications to Engineering and Science*. John Wiley and Sons, New York.

McCullagh, P. and Nelder, J.A. (1989) *Generalized Linear Models*. Second Edition. Chapman and Hall, London and New York.

Meier, P. (1953). Variance of a Weighted Mean. *Biometrics* **7**, 59–73.

Mendel, G. (1866). *Versuche Uber Pflanzenhybriden*. English translation and commentary by R. A. Fisher (1965) In *Experiments in Plant Hybridisation*, (Ed.) J.H. Bennett. Oliver and Boyd, Edinburgh.

Meyer, K. (1985). Maximum Likelihood Estimation of Variance Components for a Multivariate Mixed Model with Equal Design Matrices. *Biometrics* **41**, 153–165.

Miller, J.J. (1979). Maximum Likelihood Estimation of Variance Components - A Monte Carlo Study. *Journal of Statistical Computation and Simulation* **8**, 175–190.

Milliken, G.A. and Johnson, D.E. (1992). *Analysis of Messy Data*, Vol. 1, Designed Experiments. Chapman and Hall, London and New York.

Mitra, S.K. (1972). Another look at Rao's Minque for Variance Components. *Bulletin of the International Statistical Institute* **44**, 279–283.

Montgomery, D. C (1991). *Design and Analysis of Experiments*, Third Edition, John Wiley and Sons, New York.

Moriguti, S. (1954). Confidence Limits for a Variance Component. *Statistics and Applied Research* **3**(2), 1–13.

Mosteller, F. and Tukey, J.W. (1982). Combination of results of Stated Precision: I. The Optimistic Case. *Utilitas Mathematica* **21 A**, 155–178.

Mosteller, F. and Tukey, J.W. (1984). Combination of Results of Stated Precision: II. A More Realistic Case. In *W.G. Cochran's Impact on Statistics*, P.S.R.S Rao and J. Sedransk (Eds.), 223–252. John Wiley and Sons, New York.

Myers, R.H. and Howe, R.B. (1971). On Alternative Approximate F Tests for Hypotheses Involving Variance Components. *Biometrika* **58**, 393–396.

Neyman, J., Iwaszkiewicz, K. and Kolodziejczyk, S. T. (1935). Statistical Problems in Agricultural Experimentation. *Journal of the Royal Statistical Society* **2**, 107–154.

Patterson, H.D. and Thompson, R. (1971). Recovery of Inter-block Information when Block Sizes are Unequal. *Biometrika* **58**, 545–554.

Paull, A.E. (1950). On a Preliminary Test for Pooling Mean Squares in Analysis of Variance. *Annals of Mathematical Statistics* **21**, 539–556.

Peddada, S. D. (1989). Two Non-negative Estimators for the Model with a Common Mean. *Communications in Statistics, Simulation and Computation* **18**(2), 501–512.

Plackett, R.L. (1960). Models in the Analysis of Variance (with discussion). *Journal of the Royal Statistical Society* **B22**, 195–217.

Praire, R.R. and Anderson, R.L. (1962). Optimal Designs to Estimate Variance Components and to Reduce Product Variability for Nested Classifications. North Carolina State College, Institute of Statistical Memo Series, 313, 80–81.

Pukelsheim, F. (1981). On the Existence of Unbiased Nonnegative Estimates of Variance Covariance Components. *Annals of Statistics* **9**, 293–299.

Rao, C.R. (1970). Estimating Heteroscedastic Variances in Linear Models. *Journal of the American Statistical Association* **65**, 161–172.

Rao, C.R. (1971a). Estimation of Variance and Covariance Components. *Journal of Multivariate Analysis* **1**, 257–275.

Rao, C.R. (1971b). Minimum Variance Quadratic Unbiased Estimation of Variance Components. *Journal of Multivariate Analysis* **1**, 445–456.

Rao, C.R. (1972). Estimation of Variance and Covariance Components in Linear Models. *Journal of the American Statistical Association* **67**, 112–115.

Rao, C.R. (1973). *Linear Statistical Inference and its Applications*. 2nd Ed. John Wiley and Sons.

Rao, C. R. (1974). Projections, Generalized Inverse and the BLUEs. *Journal of the Royal Statistical Society* **B 36**, 442–448.

Rao, C. R. (1975). Simultaneous estimation of Parameters in Different Linear Models and Applications to Biometric Problems. *Biometrics* **31**, 545–554.

Rao, C.R. (1979). MINQE Theory and its Relation to ML and MML Estimation of Variance Components. *Sankhya* **B 41**(3 and 4), 138–153.

Rao, C.R. (1984). Optimization of Functions of Matrices with Applications to Statistical Problems. In *W. G. Cochran's Impact on Statistics*, P.S.R.S. Rao and J. Sedransk. (Eds.), 191–202. John Wiley and Sons, New York.

Rao, C.R and Kleffé, J. (1988). *Estimation of Variance Components and Applications*. North Holland Publishing Co.

Rao, J.N.K. (1973). On the Estimation of Heteroscedastic Varinaces . *Biometrics* **29**, 11–24.

Rao, J.N.K. and Subrahmaniam, K. (1971). Combining Independent Estimators and Estimation in Linear Regression with Unequal Variances. *Biometrics* **27**, 971–990.

Rao, P.S.R.S. (1977). Theory of the MINQUE – A Review. *Sankhya* **B 39**, 201–210.

Rao, P.S.R.S. (1982). Use of Prior Information For Estimating the Variance Components. *Communication in Statistics* **A 11**, 1659–1669.

Rao, P.S.R.S (1984). Cochran's Contributions to Variance Components Models for Combining Estimates. In *W.G. Cochran's Impact on Statistics*, P.S.R.S. Rao and J. Sedransk (Eds.), 203–222. John Wiley and Sons, New York.

Rao, P.S.R.S. and Chaubey, Y.P. (1978). Three Modifications of the Principle of the MINQUE. *Communications in Statistics* **A 7**, 767–778.

Rao, P.S.R.S and Heckler, C.E. (1997). Estimators for the Threefold Nested Random Effects Model. *Journal of Statistical Planning and Inference*, to appear.

Rao, P.S.R.S., Kaplan, J. and Cochran, W.G. (1981). Estimators for the One-way Random Effects Model with Unequal Variances. *Journal of the American Statistical Association* **76**, 89–97.

Rao, P.S.R. S and Kuranchie, P. (1988). Variance Components of the Linear Regression Model with a Random Intercept. *Communications in Statistics*, Special Issue, **A 17**(4), 1011–1026.

Rao, P.S.R.S and Miyawaki, N. (1989). Estimation Procedures for Unbalanced One-way Variance Components Model. *Bulletin of the International Statistical Institute* **2**, 243–244.

Rao, P.S.R.S. and Shimizu, I.M. (1991). Combing Estimates from Surveys. *Survey Methodology* **17**, 131–141.

Rao, P.S.R.S. and Sylvestre, E.A. (1984). ANOVA and MINQUE

Type of Estimators for the One-way Random Effects Model. *Communications in Statistics* **A 13**, 1667–1673.

Reinsel, G. (1982). Multivariate Repeated-Measurement or Growth Curve Models with Multivariate Random-Effects Covariance Structure. *Journal of the American Statistical Association* **77**, 190–203.

Reinsel, G. (1984). Estimation and Prediction in a Multivariate Random Effects Generalized Linear Model. *Journal of the American Statistical Association* **79**, 406–414.

Rosner , B. (1982). Statistical Methods in Ophthalmology: An Adjustment for the Intraclass Correlation Between Eyes. *Biometrics* **38**, 105–114.

Rudemo, M., Ruppert, D. and Streibig, J.C. (1989). Random-effect Models in Nonlinear Regression with Applications to Bioassay. *Biometrics* **45**, 349–362.

Russell, T.S. and Bradley , R.A. (1958). One-way Variances in the Two-way Classification. *Biometrika* **45**, 111–129.

Sahai, H. (1974). Simultaneous Confidence Intervals for Variance Components in some Balanced Random Effects Models. *Sankhya* **B 36**, 278–287.

Sahai, H. (1976). A Comparison of Estimators of Variance Components in the Balanced Three Stage Nested Random Effects Model using Mean Square Error as the Criterion. *Journal of the American Statistical Association* **71**, 435–444.

Sahai, H. (1979). A Bibliography on Variance Components. *International Statistical Review* **47**, 177–222.

Sahai, H., Khuri, A.I. and Kapadia, C.H. (1985). A Second Bibliography on Variance Components. *Communications in Statistics* **A14**, 63–115.

Satterthwaite, F.E., (1946). An Approximate Distribution of Estimates of Variance Components. *Biometrical Bulletin* **2**, 110–114.

Scheffé, H. (1956). Alternative Models for the Analysis of Variance.

Annals of Mathematical Statistics **27**, 251–271.

Scheffé, H. (1959). *The Analysis of Variance*. John Wiley and Sons, New York.

Searle, S. R. Casellla, G. and McCulloch, C.E. (1992). *Variance Components*. John Wiley and Sons, New York.

Seely, J. (1977). Minimal Complete Statistics for Multivariate Normal Families. *Sankhya* **A 39**, 170–185.

Seely, J.F. and Lee, Y. (1994). A note on the Satterthwaite Confidence Interval for a Variance. *Communications in Statistics* **A 23**(3), 859–869.

Shah, K.R. and Puri, S.C. (1976). Application of MINQUE Procedure to Block Designs. *Communications in Statistics* **A 5**, 191–196.

Smith C.A.B. (1980). Estimating Genetic Correlations. *Annals of Human Genetics* **43**, 265–284.

Smith, H.F. (1936). The Problem of Comparing the Results of two Experiments with Unequal Errors. *Journal of Scientific and Industrial Research* **9**, 211–212.

Smith, H.F. (1951). Analysis of Variance with Unequal but Proportional Numbers of Observations in the Sub-classes of a Two-way Classification. *Biometrics* **7**, 70–74.

Snedecor, G.W. and Cochran, W.G. (1967). *Statistical Methods*, Seventh Edition. Iowa State University Press, Ames, Iowa.

Snee, R.D. (1983). Graphical Analysis of Process Variation Studies. *Journal of Quality Technology* **15(2)**, 76–88.

Solomon, P. J. and Cox, D. R. (1992). Nonlinear Components of Variance Models. *Biometrika* **79**(1), 1–11.

Stein, M.L. (1987). Minimum Norm Quadratic Estimation of Spatial Variograms. *Journal of the American Statistical Association* **82**, 765–772.

Stram, D. O. (1996). Meta-Analysis of Published data Using a

Linear Mixed-Effects Model. *Biometrics* **52**(2), 536–544.

Sukhatme. P.V. and Narain, P. (1984).The Genetic Significance of Intra-Individual Variation in Energy Requirement. In *W.G. Cochran's Impact on Statistics*, P.S.R.S Rao and J. Sedransk (Eds.), 275–284. John Wiley and Sons, New York.

Swallow, W.H. (1981). Variances of Locally Minimum Variance Quadratic Unbiased Estimators (MIVQUEs) of Variance Components. *Technometrics* **23**, 271–283.

Swallow, W.H. and Monahan, J.F. (1984). Monte Carlo Comparison of ANOVA, MINQUE, REML and ML Estimators of Variance Components. *Technometrics* **26**, 47–57.

Swallow, W.H. and Searle, S.R. (1978). Minimum Variance Quadratic Estimation (MIVQUE) of Variance Components. *Technometrics* **20**, 265–272.

Thomas, J.D. and Hultquist, R.A. (1978). Interval Estimation for the Unbalanced case of the One-way Random Effects Model. *Annals of Statistics* **6**, 582–587.

Thompson, W.A. Jr. (1962). The Problem of Negative Estimates of Variance Components. *Annals of Mathematical Statistics* **33**, 273–289.

Tippett, L.H.C. (1931). *The Methods of Statistics*, 1st Ed. William and Norgate, London.

Trout, J. R. (1985). Design and Analysis of Experiments to Estimate Components of Variation: Two Case Studies. In *Experiments in Industry*, American Society for Quality Control, Milwaukee, Wisconsin.

Tukey, J.W. (1951). Components in Regression. *Biometrics* **7**, 33–69.

Tukey, J.W. (1949). One Degree of Freedom for Non-Additivity. *Biometrics* **5**, 232–242.

Venables, W. and James, A.T. (1978). Interval Estimates of Variance Components. *Canadian Journal of Statistics* **6**, 103–111.

Verdooren, L.R. (1988). Least Squares Estimators and Non-Negative Estimators of Variance Components. *Communications in Statistics, Theory and Methods* **A 17**, 1027–1051.

von Rosen, D. (1991). The Growth Curve Model: A Review. *Communications in Statistics, Theory and Methods* **A 20**, 2791–2822.

Vonesh, E.F. and Carter, R.L. (1992). Mixed-Effects Nonlinear Regression for Repeated Measures. *Biometrics* **48**(1), 1–18.

Wang, C.M. and Graybill, F.A. (1981). Confidence Intervals on a Ratio of Variances in the Two-factor Nested Components of Variance Model. *Communications in Statistics* **A 10**, 1357–1368.

Welch, B.L. (1951). On the Comparison of Several Mean Values – An Alternative Approach. *Biometrika* **38**, 330–336.

Westfall, P. H. (1987). Computable MINQUE-type Estimates of Variance Components. *Journal of the American Statistical Association* **82**, 586–589.

Wilk, M.B. and Kempthorne, O. (1955). Fixed, Mixed and Random Models. *Journal of the American Statistical Association* **50**, 1144–1167.

Williams, J. S. (1962) Confidence Limits for Variance Components. *Biometrika* **49**, 278–281.

Winsor, C.P. and Clarke, G.L. (1940). Statistical Study of Variation in the Catch of Plankton nets. *Journal of Marine Research* **3**, 1–34.

Wirkowski, J.J. (1975). Unbalanced Regression Analysis with Residuals having a Covariance Structure of Intra-class form. *Biometrics* **31**, 611–618.

Wolfinger, R. D. (1993). Covariance Structure Selection in General Mixed Models. *Communications in Statistics, Simulation and Computation* **22**, 1079–1106.

Yates, F. (1934). The Analysis of Multiple Classifications with Unequal Numbers in the Different Classes. *Journal of the American Statistical Association* **29**, 51–66.

Yates, F. (1940). The Recovery of Interblock Information in BIB Design. *Annals of Eugenics* **10**, 317–325.

Yates, F. and Cochran, W.G. (1938). The Analysis of Groups of Experiments. *Journal of Agricultural Sciences* **28**, 556–580.

Yates, F.and Zacopaney, I. (1935). The Estimation of Efficiency of Sampling with Special Reference to Sampling for Yield in Serial Experiments. *Journal of Agricultural Sciences* **25**, 545–577.

American Hospital Association Guide to Health Care, 1989 figures, Chicago.

Environmental Monitoring at Love Canal, Vol.III. U.S. Environmental Protection Agency (1982); Washington, D.C.

International Ozone Panel Trends Report (1990). World Meteriological Organization Global Ozone Research and Monitoring Project, NASA.

Author Index

Subject Index